Accumulation

From food punnets to credit cards, plastic facilitates every part of our daily lives. It has become central to processes of contemporary sociomaterial living. Universalized and abstracted, it is often treated as the passive object of political deliberations, or a problematic material demanding human management, but in what ways might a 'politics of plastics' deal with both its specific manifestation in particular artefacts and events, and its complex dispersed heterogeneity?

Accumulation explores the vitality and complexity of plastic. This interdisciplinary collection focuses on how the presence and recalcitrance of plastic reveal the relational exchanges across human and synthetic materialities. It captures multiplicity by engaging with the processual materialities or *plasticity* of plastic. Through a series of themed essays on plastic materialities, plastic economies, plastic bodies and new articulations of plastic, the editors and chapter authors examine specific aspects of plastic in action. How are multiple plastic realities enacted? What are their effects?

This book will be of interest to students and scholars of sociology, human and cultural geography, environmental studies, consumption studies, science and technology studies, design, and political theory.

Jennifer Gabrys is Senior Lecturer in Sociology at Goldsmiths, University of London, and Principal Investigator on the ERC-funded project 'Citizen sensing and environmental practice'.

Gay Hawkins is a Professorial Research Fellow in social and cultural theory and Director of the Centre for Critical and Cultural Studies at the University of Queensland, Brisbane, Australia.

Mike Michael is Professor of Sociology and Social Policy at the University of Sydney.

Culture, Economy and the Social

A new series from CRESC – the ESRC Centre for Research on Socio-cultural Change

Editors

Editorial Advisory Board

The *Culture, Economy and the Social* series is committed to innovative contemporary, comparative and historical work on the relations between social, cultural and economic change. It publishes empirically based research that is theoretically informed, that critically examines the ways in which social, cultural and economic change is framed and made visible, and that is attentive to perspectives that tend to be ignored or side-lined by grand theorizing or epochal accounts of social change. The series addresses the diverse manifestations of contemporary capitalism, and considers the various ways in which 'the social', 'the cultural' and 'the economic' are apprehended as tangible sites of value and practice. It is explicitly comparative, publishing books that work across disciplinary perspectives, cross-culturally, or across different historical periods.

The series is actively engaged in the analysis of the different theoretical traditions that have contributed to the development of the 'cultural turn' with a view to clarifying where these approaches converge and where they diverge on a particular issue. It is equally concerned to explore the new critical agendas emerging from current critiques of the cultural turn: those associated with the descriptive turn, for example. Our commitment to interdisciplinarity thus aims at enriching theoretical and methodological discussion, building awareness of the common ground that has emerged in the past decade, and thinking through what is at stake in those approaches that resist integration to a common analytical model.

Series titles include:

The Media and Social Theory (2008)
Edited by David Hesmondhalgh and Jason Toynbee

Culture, Class, Distinction (2009)
Tony Bennett, Mike Savage, Elizabeth Bortolaia Silva, Alan Warde, Modesto Gayo-Cal and David Wright

Material Powers (2010)
Edited by Tony Bennett and Patrick Joyce

The Social after Gabriel Tarde: Debates and assessments (2010)
Edited by Matei Candea

Cultural Analysis and Bourdieu's Legacy (2010)
Edited by Elizabeth Silva and Alan Ward

Milk, Modernity and the Making of the Human (2010)
Richie Nimmo

Creative Labour: Media work in three cultural industries (2010)
Edited by David Hesmondhalgh and Sarah Baker

Migrating Music (2011)
Edited by Jason Toynbee and Byron Dueck

Sport and the Transformation of Modern Europe: States, media and markets 1950–2010 (2011)
Edited by Alan Tomlinson, Christopher Young and Richard Holt

Inventive Methods: The happening of the social (2012)
Edited by Celia Lury and Nina Wakeford

Understanding Sport: A socio-cultural analysis (2012)
John Horne, Alan Tomlinson, Garry Whannel and Kath Woodward

Shanghai Expo: An international forum on the future of cities (2012)
Edited by Tim Winter

Diasporas and Diplomacy: Cosmopolitan contact zones at the BBC World Service (1932–2012)
Edited by Marie Gillespie and Alban Webb

Making Culture, Changing Society
Tony Bennett

Interdisciplinarity: Reconfigurations of the social and natural sciences
Edited by Andrew Barry and Georgina Born

Objects and Materials: A Routledge Companion
Edited by Penny Harvey, Eleanor Conlin Casella, Gillian Evans, Hannah Knox, Christine McLean, Elizabeth B. Silva, Nicholas Thoburn and Kath Woodward

Accumulation: The material politics of plastic
Edited by Gay Hawkins, Jennifer Gabrys and Mike Michael

Theorizing Cultural Work: Labour, continuity and change in the cultural and creative industries
Edited by Mark Banks, Rosalind Gill and Stephanie Taylor

Rio de Janeiro: Urban life through the eyes of the city (forthcoming)
Beatriz Jaguaribe

Devising Consumption: Cultural economies of insurance, credit and spending (forthcoming)
Liz Mcfall

Unbecoming Things: Mutable objects and the politics of waste (forthcoming)
Nicky Gregson and Mike Crang

E·S·R·C ECONOMIC & SOCIAL RESEARCH COUNCIL

CRESC Centre for Research on Socio-Cultural Change

Accumulation

The material politics of plastic

Edited by
Jennifer Gabrys, Gay Hawkins and Mike Michael

Routledge
Taylor & Francis Group
LONDON AND NEW YORK

First published 2013
by Routledge
2 Park Square, Milton Park, Abingdon, Oxfordshire OX14 4RN

Simultaneously published in the USA and Canada
by Routledge
711 Third Avenue, New York, NY 10017

First issued in paperback 2017

Routledge is an imprint of the Taylor & Francis Group, an informa business

British Library Cataloguing in Publication Data
A catalogue record for this book is available from the British Library

Library of Congress Cataloging in Publication Data
Accumulation : the material politics of plastic / edited by Jennifer Gabrys, Gay Hawkins, and Mike Michael.
p. cm. – (Culture, economy and the social)
Includes bibliographical references and index.
1. Plastics–Material aspects.. 2. Plastics–Social aspects. 3. Plastics–Political aspects. I. Gabrys, Jennifer, editor of compilation. II. Hawkins, Gay, editor of compilation. III. Michael, Mike, editor of compilation.
HC79.S3A25 2013
338.4'76684–dc23
2012050792

ISBN 13: 978-1-138-06306-8 (pbk)
ISBN 13: 978-0-415-62582-1 (hbk)

Typeset in Times New Roman
by Taylor & Francis Books

Contents

Illustrations

Figures

Contributors

Mandy Barker is a photographic artist working in the United Kingdom. Since graduating from the MA Photography at De Montfort University (2011), her work involving plastic marine debris has received international recognition. Her series of work, 'SOUP', has been published in over 15 countries, and has circulated in *Time* magazine and on CNN. As part of the Blurb Photography Book Now Award, her images were projected on to the walls of The Aperture Foundation in New York. Selections of her work have been made by Stephanie Braun of The Photographer's Gallery London and for *Source* magazine. Francis Hodgson of the *Financial Times* nominated Mandy's work for the fourth cycle of the Prix Pictet Award 2012, which is the world's leading photographic award in sustainability. In June 2012, Mandy joined scientists aboard a Plastic Research Expedition sailing from Japan to Hawai'i through the Tsunami Debris Field in the Pacific Ocean. Her objective was to continue her work at source and from a location not previously attempted, alongside scientists, in order to generate a deeper understanding of the detrimental effects of plastic on marine life.

Bernadette Bensaude Vincent is a philosopher and historian of science. She is currently a Professor in the Department of Philosophy at Université Paris I Panthéon-Sorbonne and a member of the Institut universitaire de France. Her research topics span from the history and philosophy of chemistry to materials science and nanotechnology, with a continuous interest in science and public issues. Among her recent publications are *La science et l'opinion. Histoire d'un divorce* (2003); *Se libérer de la matière? Fantasmes autour des nouvelles technologies* (2004); *Chemistry: The Impure Science* (coll. Jonathan Simon, 2008); *Matière à penser* (2008); and *Les Vertiges de la technocience: Façonner le monde atome par atome* (2009).

Joe Deville is a researcher based at the Centre for the Study of Invention and Social Process at Goldsmiths, University of London. He is currently writing a book looking at consumer credit default and collection, drawing on research from his PhD thesis, completed in 2011. It will follow the changing calculative landscapes through which heavily indebted and defaulting consumer credit borrowers move, from periods of borrowing, to managing debts, to being confronted by debt collectors. It also focuses on the problematic of debt default from the point of view of the collector, exploring the increasingly sophisticated techniques being used to attempt to convince debtors to repay.

Tom Fisher is Professor of Art and Design at Nottingham Trent University, United Kingdom. A graduate in Fine Art, he has worked as a designer and maker of furniture, and wrote his PhD in the Sociology Department at the University of York, concentrating on everyday experiences of plastic materials. His current research focuses on the materiality of human-object relationships and their implications for sustainability. In this, he draws on his background as a maker and on perspectives from the sociology of consumption. He recently published *Designing for Re-Use: The Life of Consumer Packaging* (Earthscan, 2009), which reports on ethnographic work in the United Kingdom that uncovered the range of ways in which people reuse packaging. The book presents a number of ways of thinking about the phenomenon of packaging reuse, and indicates the role that design can play in promoting it.

Jennifer Gabrys is Senior Lecturer in Sociology at Goldsmiths, University of London. Her research investigates environments, material processes and communication technologies through theoretical and practice-based work. Projects within this area include *Digital Rubbish: A Natural History of Electronics* (University of Michigan Press, 2011), which examines the materialities of electronic waste, and a study currently underway on environmental sensor technologies and practices, titled *Program Earth: Environment as Experiment in Sensing Technology.*

Gay Hawkins is a Professorial Research Fellow in social and cultural theory and Director of the Centre for Critical and Cultural Studies at the University of Queensland, Brisbane, Australia. She is currently completing a book on the rise and impacts of bottled water called *Plastic Water*. Previous publications include *The Ethics of Waste* (Rowman and Littlefield, 2005); and 'Plastic Materialities' in Braun and Whatmore (eds), *Political Matter* (University of Minnesota Press, 2010). In 2013 she is commencing a major study called *The Skin of Commerce*, investigating the pragmatics and performativity of plastic food packaging: how plastic became a market device in post-war food economies, how it contributed to major urban waste problems, and how it has become a political material.

Max Liboiron is a postdoctoral fellow in Media, Culture and Communication at New York University, where she researches theories of scale in relation to environmental change. Her dissertation, 'Redefining Pollution: Plastics in the Wild', investigates scientific and advocate techniques to define plastic pollution, given that plastics are challenging centuries-old concepts of pollution as well as norms of pollution control, environmental advocacy and theories of contamination. Her work has been published in *eTOPIA:*

Canadian Journal of Cultural Studies, *Social Movement Studies: Journal of Social, Cultural and Political Protest* and in the *Encyclopedia of Consumption and Waste: The Social Science of Garbage*. She writes for the Discard Studies Blog (discardstudies.wordpress.com), and is a trash artist and environmental activist. See www.maxliboiron.com.

James Marriott and **Mika Minio-Paluello** are part of Platform, a London-based arts, human rights and environmental justice organization that has pioneered diverse and far-reaching responses to issues of international importance for more than 25 years. Based on core values of solidarity, creativity and democracy, Platform combines art, activism, education and research to achieve long-term systemic goals. Since the mid-1990s, the group has had a strong focus on the international oil and gas industry – researching, campaigning against its impacts and creating artworks in a wide range of media. Together, they have recently published *The Oil Road: Journeys from the Caspian Sea to the City of London* (Verso, 2012). See www.platformlondon.org.

Mike Michael is Professor of Sociology and Social Policy at the University of Sydney. He is a sociologist of science and technology with research interests in the relation of everyday life to technoscience, biotechnological and biomedical innovation and culture, and process methodology in the social sciences. Recent research projects include an examination of the ethical aspects of HIV pre-exposure prophylaxis (with Marsha Rosengarten) and the exploration of energy demand reduction through sociological and speculative design techniques (with Bill Gaver and Jennifer Gabrys).

Jody A. Roberts is the Director of the Center for Contemporary History and Policy at the Chemical Heritage Foundation in Philadelphia, PA, United States. Roberts's work explores the intersections of emerging molecular sciences and public policy and the ways in which tensions brought about between the two get resolved. He received advanced degrees in science and technology studies from Virginia Tech, where he cultivated an interest in the practice of the molecular sciences and the ways in which they are shaped by internal architecture and design (e.g. technologies of the laboratory) and the politics of the broader world (e.g. chemical regulations). He also holds a Bachelor of Science degree in chemistry from Saint Vincent College. Roberts held fellowships at CHF before joining the staff of the Center for Contemporary History and Policy in 2007. He lectures in the History and Sociology of Science Department at the University of Pennsylvania and the Centre for Science, Technology, and Society at Drexel University. He is also a Senior Fellow in the Environmental Leadership Program.

Shige Takada received his PhD from Tokyo Metropolitan University in 1989. His speciality is trace analysis of organic micropollutants. The target compounds include persistent organic pollutants (POPs; e.g. PCBs, DDTs, PBDEs, PAHs), endocrine-disrupting chemicals (e.g. nonylphenol, bisphenol A), pharmaceuticals, as well as anthropogenic molecular markers. His research field encompasses Tokyo Bay and its vicinities, to South-East Asia, to Africa. In 2005, Shige Takada started International Pellet Watch, the global monitoring of POPs by using beached plastic resin pellets (www.pelletwatch.org). He has been working with ~50 non-governmental organizations and individuals who have concerns about marine plastics pollution. Shige Takada (Hideshige Takada) has authored more than 100 peer-reviewed papers in international journals and has attended more than 20 invited conferences.

Richard C. Thompson is Professor of Marine Biology at Plymouth University, United Kingdom. He is a marine biologist specializing in the ecology of shallow-water habitats. He has a first-class degree in Marine Biology from the University of Newcastle upon Tyne and a PhD from the University of Liverpool. He moved to Plymouth University in 2001, where he helps run the BSc Marine Biology degree programme and lectures in marine ecology. Much of his work over the last decade has focused on marine debris. In 2004, his group reported on the presence of microplastics in the environment in the journal *Science*. Subsequent research examined the extent to which microplastics were retained upon ingestion by marine organisms, and the potential for microplastics to transport persistent organic pollutants to organisms. In 2007, he was invited to be lead guest editor for a 200-page volume of the *Philosophical Transactions of the Royal Society B*, focusing on plastics, the environment and human health. He is a co-author of the European Union Marine Strategy Framework Directive Task Group 10 on marine litter, and recently has prepared two reports on marine debris for the United Nations Global Environment Facility: *Marine Debris as a Global Environmental Problem: Introducing a Solutions Based Framework Focused on Plastic* (2011); and *Impacts of Marine Debris on Biodiversity: Current Status and Potential Solutions* (2012).

Andrea Westermann works in the History Department of the University of Zurich, specializing in the history of science and technology and environmental history. She wrote her first book on the history of plastics in West Germany. She is currently writing a book on knowledge validation in late nineteenth- and early twentieth-century geology. Recent publications include 'Disciplining the earth: earthquake observation in Switzerland and Germany circa 1900', in *Environment and History*, and 'Inherited territories: the Glarus Alps, knowledge validation, and the genealogical organization of nineteenth-century Swiss alpine geognosy', in *Science in Context*. She is partner investigator with Gay Hawkins on the research study, *The Skin of Commerce*.

Introduction

From materiality to plasticity

Jennifer Gabrys, Gay Hawkins and Mike Michael

Figure I.1 Academics and plastic at work

In Figure I.1, two of the editors are busily working on the introduction to this volume, moments after we had decided on how to approach it. It should be obvious that we are surrounded by plastic. There are plastic objects such as pens and computers, the plastic packaging of a water bottle and a punnet of cherries; there is plastic covering the chairs, laminating the table and encasing the printer; and there is plastic making the meeting physically feasible – plastic in the light fittings, in the carpets, in the window frames (it was a cold, dank London June day). Inevitably, there is plastic in our clothes and on our bodies. This is our first meeting together since the *Accumulation* conference, an event we had co-organized the year before that brought together a number of the authors included in this collection to examine the material force of plastic as raw material, object and process. Organizing this interdisciplinary event and collection of speakers would have been impossible without a wealth of other plastics, from the plastic in trains and airplanes to the plastic in credit cards. All this simply goes to underscore how central plastic has become to processes of contemporary sociomaterial living.

This is an easy observation to make. From the first fully *synthetic* plastic – Bakelite, developed in 1907 – to the current proliferation of polymers, we produce, consume and dispose of plastics in untold quantities. In the corresponding plethora of studies on the rise of plastic, its rapid uptake and ubiquity are regularly noted. Previous research on plastics seems to have appeared every decade or so since its widespread application within post-World War II consumer economies. As early as 1957, cultural critic and semiotician Roland Barthes wrote a classic essay on plastic, which he developed as a comment and critique of this substance as he witnessed it arrayed in a plastics exhibition in Paris. In 1986, designer Ezio Manzini published *The Material of Invention*, which emphasized the material performativity and interchangeability of plastics. In 1995, historian Jeffrey Meikle wrote a sophisticated and nuanced account of how plastics had influenced American culture. More recently, an increasing number of pop-culture and pop-science commentaries on plastics are emerging that chart the toxicity, intractability and spread of plastics. Susan Freinkel's (2011) *Plastic: A Toxic Love Story* exemplifies this trend by mixing due recognition of the benefits and necessity of plastic with a consumer guide to its environmental and noxious horrors. In different ways, these texts explore how *things* have become decidedly synthetic to the point where plastic now appears as the archetypal material of invention, mass consumption and ecological contamination.

Beyond this opening vignette, and gesturing toward post-war and recent literature, a host of other plastics-related issues are waiting to be excavated. The purpose of this collection is to explore the vitality, complexity and irony of plastics, and to examine a range of plastics-related issues that cut across art and design practices, humanities, natural sciences, politics and the social sciences. We are not seeking to establish a general narrative about the evolution of plastics. Nor do we wish to frame plastic as emblematic of social and environmental change. Rather, the aim is to capture the multiplicity and complexity of plastic by engaging with its processual materialities, or *plasticity*. As we suggest here, plasticity extends not just to the multiple forms and uses of plastics, but also includes the ways in which plastics are integral to contemporary material processes, and even give rise to events such as environmental or bodily accumulation that present unexpected and often undesirable modes of material transformation.

Accumulation engages with the particularity of plastics in order to draw out these aspects and implications of plasticity. The collection presents a series of chapters that address plastic in its concrete manifestations, including PET (polyethylene terephthalate) water bottles, credit cards, degrading and biodegradable plastics, everyday litter, marine debris, rapid prototyping, mobile phones, oil and oil transportation. These examples provide richly detailed accounts of the ways in which plastic is woven into and enacted through social, cultural, political, technoscientific, ecological and economic practices. In all of these accounts, plastics are part of transformative material engagements. What emerges through these empirical object studies is how plasticity provides particular ways of thinking about and advancing understandings of materiality *as process*.

Both the fact that plastic has become an object of variegated analysis more generally, and that this collection has become possible to assemble, deserve comment. On the one hand, we can put this down to a number of developments in the social sciences – for instance, the recent 'turn to the object' exemplified in such perspectives as material culture studies, science and technology studies (STS), or recent versions of the sociology of everyday life. These approaches have been particularly useful in helping us to engage with plastic materials and objects, not least by rendering them 'seeable' and lifting them from their fog of familiarity or background passivity, thereby making them interestingly and productively unfamiliar. Plastic complicates this turn, however, for it is not just the world of objects that is defamiliarized, but also the material properties that constitute those objects. Plastic draws attention to the materiality of objects *and* the shifting properties of those materials.

Plastics have also become defamiliarized by making their presence felt, by becoming a very insistent matter of concern, where discarded plastics and a whole range of plasticizers added to polymers increasingly are understood to induce harmful effects on bodies and environments. More than any other material, plastic has become emblematic of economies of abundance *and* ecological destruction. If the post-war 'Plastic Age' was cleaner and brighter than all that preceded it, this boosterism has now become intertwined with significant anxiety as the burden of accumulating plastic waste registers in environments and bodies. The indeterminate and harmful materialities of plastic are now surfacing and demanding urgent attention. Over the last 10 years, there has been increasing public controversy about the endocrine-disrupting effects of plastic, about the emergence of massive plastic gyres in several oceans, and about the ethical and environmental impacts of the global spread of disposable plastic cultures. Because they are made in part from petroleum, plastics have become a marker of dwindling natural resources and accumulating synthetic pollution, with their limited degradability signalling indefinite processes of environmental degradation. Plastics simply refuse to go away, and their material recalcitrance forces us to acknowledge the ways in which plastics persist long after their use value is exhausted (Gabrys 2011).

This edited collection then engages with these multiple qualities of plastic to think through questions related to emerging areas of materialities research. There are a growing number of studies on the ways in which materials matter and inform political engagements (Bennett 2010; Braun and Whatmore 2010;

Hawkins 2010). We situate this volume in conversation with these studies, and through these investigations into plastics consider how plasticity distinctly informs our ways of encountering and mobilizing material research. Among the questions to be addressed in this collection, we ask how it may be possible to engage with the processual materiality or plasticity of plastic without fixing it as an object of study or illustrative case. In the various chapters presented here, we ask how the recalcitrance of plastic, its durability and persistence, might reveal the material and relational exchanges that take place between humans and non-humans. We also consider further how recognition of the material force of plastics prompts new forms of politics, environmental responsibility and citizenship. A key concern running through these questions is how we might begin to develop an analytics attentive to plastics in order to provoke invention and invite new forms of material thinking. This, as we suggest in this introduction and throughout this collection, involves attending to the *material politics* that emerge through the processual materialities of plastics.

Material politics and the event of plastics

What stands out from the photograph with which we began this introduction is the sheer number of plastic objects and materials that surround the 'human subjects'. There are now over 10,000 types of plastic polymers in use, and worldwide consumption of plastic has gone from barely measurable quantities in 1940 to 260 million tons per annum today (Thompson *et al.* 2009). As such, we might ask how we should think about the multiplicity of plastics that make themselves present in the event of this meeting and beyond. Is there a plurality of different plastics and their related objects – some of which feature overtly or covertly in this scene? Or is there a singular family of plastics the common, unique and abiding property of which is the capacity to take on multiple forms? This raises the issue of the ontological status of plastic. Is it something that has certain intrinsic properties or recalcitrance, or does plastic emerge from the multiple relations in which it is embroiled: in chemistry, in industry, in processes of marketing and consumption, in plastics waste management? Or is this dichotomy itself less than useful? Put another way, can we move beyond this staging of the issues by, rather than *applying* this dichotomy (and its problematization) to plastic, thinking it *through* the specificities of plastics. In this way, the focus shifts to developing an empirical ontology of how multiple plastic realities are enacted and their effects.

What is plastic doing in the world? What might it do? Questions about the concrete effects of specific manifestations of plastics quickly lead to political entanglements, but the political questions that emerge in this study do not just stem from a human assessment of 'bad impacts'. Instead, we suggest that plastics generate a series of causes or political reverberations that genuinely constitute modes of material politics, which emerge from the concrete events of plastics in the world. The material politics of plastics can then be seen as emergent and contingent, where plastics set in motion relations between things that become sites of responsibility and effect. From this perspective, a material politics informed by plastics is less oriented toward asserting that materials are always already political. Instead, in this collective study on plastics, we variously focus on when and how plastics as materials *become* political. Through which material processes and entanglements do plastics 'force thought' and give shape to political concerns (Stengers 2010)?

Plastics, as it turns out, may teach us something about politics. The mutability of this material generates distinct political encounters and events. Rather than assume the prior political status of materials, we then ask in relation to plastics: in what ways are these materials political, and how do political moments and matters of concern emerge in relation to these materials? This point is important, since from this perspective we begin to think about the new relationalities that plastic is generating, and how these relationalities become sites of responsibility. Rather than argue for the simple elimination of plastics, we suggest that the material politics of plastics require that we attend to these unfolding relationalities and responsibilities in order to ask: to which material and political futures are we committing ourselves, and in what ways might an inattention to the material politics of plastic foreclose opportunities for inventing different material futures?

This collection seeks to offer insight into the ways that a 'politics of plastics' must deal with both its specific manifestation in particular artefacts and events, and its complex dispersed heterogeneity. This approach challenges the abstraction and universalization of plastic as the passive object of political deliberations or a problematic material demanding human management. Instead, the focus is on how plastics rework material relations within and through synthetic and dynamic processes. Plastics, in this sense, can refresh thinking on materiality by forcing an attention to material processes within which we are specifically situated, but which are not bounded objects of study.

As the plastics photograph and this discussion suggest, the *event* of plastics is a key way in which the materiality of plastics is encountered and understood. Even to talk about the photograph that frames this introduction as a representation of plastics is to neglect the fact that it is *doing* something. Clearly, by foregrounding certain elements of plastic and the academic approach to the study of plastic while downplaying others, the photo is setting up – performing, enacting – a very particular vision of plastic and academic work. This vision can have effects – at the very least, to persuade readers that they should subscribe to that vision (and change their practice accordingly). However, the performative dimension of plastic resides not simply in our enactment of it through a photograph; plastic is enacted more broadly through the complex relations that comprise it.

Framing the photograph in a different way, we can see before us an event in which a series of elements come together – from the sub-atomic to the social-relational – to render plastic in a range of ways. We have touched on some of these already: plastic tools, materials, components, markers of resource depletion and environmental degradation, objects of academic study. However, out of the event emerge not only the non-human elements, but also the human. In other words, we also see the array of plastic partially enacting the academics: in the specificity of the event (we might also call this an assemblage), these plastics (along with myriad other elements) enable particular sorts of humans to emerge.

It will not have escaped notice that to talk about enactment, emergence and indeed affect is also tacitly to raise the matter of causality. Causality underpins much of the assessment of the environmental impact of plastics, but this is not necessarily a linear causality. Often the impact of plastics operates within complex systems in which it is not at all clear what the outcomes will be. That is to say, in this collection of plastics studies we attend to the 'emergent causality' of plastics (Bennett 2010), and how this is at play in both the production of environmental impacts and the sociomaterial enactments of humans – and indeed, more-than-humans.

This last phrase – the more-than-human – also throws into relief the fact that our initial reading of the photograph may be disingenuous on another level. We have been seeing plastics and humans separately, as distinct entities interacting with one another. Yet this opening commentary has in many ways focused mainly on their intra-actions: their mutual emergence, their becoming-with. Or, to frame it another way – which is somewhat less concerned with the standard units of human and non-human – what we can also see are concrescences (here drawing on Whitehead 1929), at least minimally composed of combinations of academic and computer, or academic and pen. As such, when we turn to the analysis of emergent causality or enactment, we also need to ask about what might be the most fruitful units of analysis. At which level can plastic objects and events be identified, and with which sorts of entanglements are these objects and events further connected?

Material thinking as inventive problem making

Finally, and relatedly, the photograph and the taking of the photograph are, like plastic, nothing if not oriented to the future. The photograph points to the future promise of a book, just as plastic is wrapped up in a panoply of expectations, hopes, fears and hypes of plastic. However, these promissory accounts that accompany plastics are but another element in the event in which plastics emerge – they are performative of what plastic might be, and how plastic might affect the world – but there is no guarantee of realizing that particular promise, or any particular future. The event, including the event of plastic, is chronically open. This begs the question of not just what our unit of analysis should be, but also when and how it must incorporate this openness. More provocatively, what does analysis mean when our object is fundamentally in-process? Is the function of this volume to provide a solution to the problem of plastic's processuality? Somehow to pin it down? Or is it to pose the problem better, more inventively?

Such 'inventive problem making' (Fraser 2010; Michael 2012) is a strategy we have developed by bringing together diverse perspectives on plastics. In editing this volume, we have gathered together an interdisciplinary range of engagements with plastics. The chapters collected here draw on very different epistemic traditions, from cultural studies to design and the sociology of science and technology, as well as history, geography, economic sociology, organic biochemistry, environmental science and environmental politics. The immediate issue that might arise would concern the 'management' of such interdisciplinarity, but we might see such disciplinary interactions as an opportunity for producing new 'objects of knowledge' by virtue of the reconfigured relations brought together through such interactions. Here, what counts as 'object' and 'knowledge' can shift dramatically.

In this sense, we could see the interplay of two classical epistemological directions, respectively and crudely, from knowledge to the material, and from the material to knowledge. Another way of saying this would be that changes in knowledge reshape our engagement with the material world, at the same time as the material world affects our knowledge and knowledgeability. This dynamic is at the heart of 'material thinking' (Carter 2004; Thrift 2006). One of the objectives of this book is to trace how distinct types of material thinking emerge and might yet be invented in relation to the specificities of plastics. Such a move in part takes up Stengers's interest in engaging non-humans 'as causes for thinking', which, because they 'force' thought, cannot be taken for granted within a standard 'state of affairs', but must be encountered as shifting connections and new, if problematic, sites of collective becoming (Stengers 2010: 14–17).

'Ecologies of practice' across humans and non-humans, as well as disciplines, is a term that Stengers (2005) coins to describe these moments when new connections, obligations and possible modes of speculation emerge. We take up such an experiment in ecologies of practice, working not just across carbon-based humans and polymer-based non-humans and the multiple new arrangements that emerge through these intersections, but also drawing on an interdisciplinary range of approaches to plastics. Interdisciplinarity is a process in which often mutually incompatible framings are brought together, but do not always easily align.

New plastic objects and ways of encountering plastics emerge through these reconfigured relations, and participants may even change in the process of negotiating what counts as thinking with plastics. These changes are not simply cognitive or epistemic, or even ethical – they are also affective. That is to say, in interdisciplinary discussions of plastic, there is a certain human plasticity that is exercised – an affective accommodation of the other that allows for initial toleration if not always understanding, and common practical orientation if not always mutual comprehension. The point is that it will also be necessary to address how the juxtaposition of approaches that characterize this volume yields partial connections that can resource a complex reimagining (that operates on epistemic and affective registers) of how plastic can be known and, indeed, enacted.

Accumulation brings divergent perspectives into a timely dialogue around and through the multifarious materialities and events that comprise plastics. Our inventive approach to the complex material issues of plastics is to work from within the concrete events and relations that specific types of plastics enable or sustain. Through a series of themed sections on Plastic Materialities, Plastic Economies, Plastic Bodies and New Articulations of plastic, specific aspects of plastic in action are examined in the chapters that follow. These themes are both conceptual in the sense that they frame how we approach plastic and also ontological in the sense that they signal the empirical unfolding of plastics in the world. To talk of 'materialities' is to acknowledge the stuff of plastic, the physical properties of a substance and also the issue of materiality as a key focus of recent social theory. To talk of 'economies' is to recognize plastic as a locus of value both as a global industry and a specific market device. In the work of everything from packaging to the credit card, plastic co-articulates economic actions; it is both a medium of business and a material actant. To talk of 'bodies' is to focus on the ways in which we are entangled with plastic, the multiple interminglings in which humans (and non-humans) find themselves becoming with plastic, whether they like it or not. To talk of 'new articulations' is to wrestle with plastic as a material that forces thought. The problematic ontologies and matters of concern that emerge with plastics become manifest in particular artefacts and events from oil pipeline protests to ocean gyres. In this way, plastic is articulated as complex dispersed heterogeneity, not simply as bad stuff to be eliminated or avoided but as a material with the capacity, in certain settings and events, to *provoke* political actions.

Plastic materialities

What is plastic if not material? Despite all the fascination with plastic objects and their ubiquity, what about the material locked up in these things? How can we investigate the relationship between the properties of plastic as a chemical-material substance and all the things and products it becomes? This interaction between physical properties and the seemingly endless capacity of the material to be materialized is the focus in Part I of this collection. Taking up Manuel DeLanda's imperative to study the behaviour of matter in its full complexity, our aim is to understand the processes whereby plastic emerges as a distinct material and the ways in which its material properties – expected and unexpected – emerge and are enacted in objects. This is much more than an historical investigation or a venture into chemistry or materials science. As DeLanda (2005) argues, the behaviour of materials is as much a philosophical issue as an empirical one. The close observation of materials and direct interaction with their properties was the basis of the earliest philosophies of matter. Many of these early practices of material thinking explored the variability and behaviour of materials; they were sensual and metaphysical as well as empirical. This diversity was lost with the rise of specialisms and the desire to identify essential and fixed properties in materials. The rise of chemistry was symptomatic of this shift, signalling 'the complete concentration of analysis at the levels of molecules [and] an almost total disregard for higher levels of aggregation in solids' (DeLanda 2005: 2).

However, it couldn't last. As Bernadette Bensaude Vincent shows in her opening chapter, 'plastics' challenged many of the conventions of chemistry. For it was what they could do – the multiplicity of arrangements of molecules and forms that emerged – that defined them. Unlike glass or wood, this material was referred to by its properties, by its capacity for change and proliferating uses. In this way, 'plastic' and 'plasticity' were as much cultural classifications as technical ones, but these entanglements across material capacity, technical understanding and cultural uses were shifting and uneven. Initially, as Bensaude Vincent shows, this very changeability was regarded with disdain, as a sign of inferiority and cheap substitution. Plasticity was a mark of the inauthentic. However, as the post-war Plastic Age escalated, these very properties became markers of positive value. Plasticity was an indicator of 'protean adaptability' and mass accessibility; it was the material that democratized consumption. This cultural shift reflected the dramatic expansion of uses and applications as the performance of plastic was enhanced. The development of thermosetting polymers foregrounded the philosophical complexity of plastic – namely, the way in which, through processes of synthesis and shaping, matter and form emerge simultaneously.

If Bensaude Vincent documents the plasticity of plastic, Mike Michael draws attention to the limits of such plasticity – that is, through the laboratory and factory processes where plasticity comes to be realized. Plasticity thus belongs to the sphere of production – or rather, it has until very recently. With the introduction of home 3D printers, it would appear that the plasticity of production now extends into the domestic sphere, and, with this redistribution of plastic's plasticity, anyone can produce anything at any time. Michael traces how these abstract claims for the 'democratization of production' of plastic objects are rendered, and the ways in which they are intertwined with reconfigurations and retrenchments of space (domestic, industrial, environmental), human bodies and minds (manual and ICT skills) and the future (utopian, apocalyptic). In other words, domestic 3D printers mediate a range of plastic's emergent properties that are complexly sociomaterial, spatiotemporal and actual-virtual.

Plastic economies

In Part II of this collection, we investigate what plastic does in terms of generating economic value. What is the efficacy of plastic in processes of economic accumulation? How can we think about plastic as an economic agent? Here, the focus is on plastic as diverse industries *and* as distinct market devices. Our interest is in understanding the evolution of the plastics and petrochemical industries in the twentieth century, and to examine their interrelationships with governments, markets, consumers and the environment. We also investigate how particular plastic artefacts become central to the organization of exchange, practices of value and markets. How would the accumulation of surplus value happen without the ubiquitous credit card, or the retail packaging that makes commodities and consumption fluid and mobile?

In Chapter 3, Gay Hawkins explores the rise of polyethylene terephthalate, or PET as it is commonly known. Her goal is to understand how this distinct plastic, which most often takes the form of disposable single-use bottles, became 'economically informed' – that is, how it acquired the capacity to articulate new economic actions in the beverages industry. Central to her analysis is a concern with the nature of disposability. How did plastic come to acquire the character of a throw-away or single-use material? What economy of qualities was developed to enact a temporality of transience, and how did this generate troubling shadow realities such as massive increases in plastics waste? Hawkins uses a topological approach to pursue these questions. She traces how the PET bottle can be considered a conduit of topological relations that connects plastics waste with plastics production and consumption. In her analysis, the bottle is a medium by which the multiple enactments of disposability become co-present and related, showing how the ever-growing flow of plastic moves in chaotic and multiple directions. PET bottles are *made to be wasted* and their anticipated future is inscribed in their multiple presents.

If Hawkins emphasizes the role of PET in modes of disposability, Andrea Westermann focuses on the capacities of vinyl in advancing consumer democracy in West Germany from the 1930s onwards. Vinyl was a key ersatz material that enabled a proliferation of consumer goods (and wartime materials) that at once advanced German industry and contributed to individual prosperity. Westermann emphasizes the extent to which vinyl facilitated a version of consumer democracy based on consumer citizens. Vinyl effectively became a material-political medium that generated and reinforced the possibilities for individual choice and mass consumption. However, vinyl has not been without its problems, and Westermann charts how consumer citizens became citizen activists when confronted with increasing evidence of the toxicity of vinyl.

The final chapter in this section focuses on that quintessential plastic object of economic exchange: the credit card. Joe Deville charts the rise of the credit card, and maps the practices and campaigns whereby credit cards became more prevalent as a medium of exchange, and indeed enabled more plastic or fluid modes of credit and consumption. Deville considers the extent to which the plasticity of the credit card is a key part of its circulation, and extends this analysis to contemporary examples of debt and default. In these cases, the material presence of the credit card may become a site where the promise of credit is revoked through the demand that credit cards be cut up or returned, or it may become a site of protest, where credit card users refuse to comply with the material demands of credit card companies. The plasticity of credit cards, Deville suggests, is not incidental to the functioning of credit, but becomes a feature that unfolds in multiple and at times contradictory ways.

Plastic bodies

While plastic economies in many ways demonstrate the ways in which plastics have become more pervasive and central to the circulation of value, plastic bodies grapple with the ongoing effects of living in increasingly plasticized environments, as explored in Part III. Flesh and environments alike are now being reconstituted through the lingering and residual effects of plastics, but many of these effects are relatively new phenomena of study, with endocrine disruption and the concentration of persistent organic pollutants (POPs) now becoming topics of more thorough-going concern and research. Our daily and bodily exposure to plastics forms even before they circulate as residual and wasted matter in environments, and may settle at the level of habits and affective attachments as much as bodily incorporations. As Tom Fisher documents in his design-based discussion on plastic surfaces and mobile phones, the tactile engagements with touch-screen surfaces are a critical part of the processes through which we engage with mobile phones, and eventually disengage when these plastic objects begin to show signs of wear. The proximity of mobile phones and touch screens to bodies, Fisher finds, gives rise to distinct practices and material evaluations of plastics. The process of sustaining flawless plastics and disposing of them when degraded, he argues, can be highly influenced by how our physical interaction with devices unfolds through affective modes of embodiment.

The intimate and persistent ways in which we encounter plastics begin even before birth, as Jody Roberts precisely inventories in his compelling account of the delivery of his daughter, Helena, who was required to spend her first days and weeks of life in intensive care in order to receive medical attention, a process thoroughly dependent upon plastics. Roberts explores how someone who previously had counted himself a 'plastiphobe', and who had deliberately avoided the use of plastics, began to grapple with these new dependencies on plastics and all that they enabled. Drawing on a science and technology studies perspective, Roberts unflinchingly examines the ways in which we have and continue to become plastic – and the molecular, bodily and environmental plastic practices and effects with which increasingly we are entangled.

Following on from this assessment of the many ways in which we are entangled with both the seemingly indispensable and yet often harmful effects of plastics, Max Liboiron suggests that the particular behaviours of plastics and plasticizers may require that we rethink our models of pollution. Liboiron proposes a move away from an exclusively linear point-source understanding of pollution as occurring from a discrete source and moment in time, and instead suggests we reconsider the older and seemingly folkloric understanding of pollution as a miasma in order to account for the dispersed, multifarious, potentially low-level and yet persistent exposures to plastics and plasticizers in the environment. A miasma model of pollution, Liboiron suggests, may influence corresponding approaches to environmental policy and justice by focusing on environmental distributions and concentrations of harm, for instance, rather than individual exposure to individual substances.

Concluding this section on the entanglements of plastics and bodies, Richard Thompson brings a perspective from marine biology to the impacts of plastics on humans and non-humans. He draws connections across how the effects of phthalates and Bisphenol-A (BPA) on marine organisms, for instance, may also have consequences for humans. While uncertainty still persists about the effects of many plastics and plasticizers, harmful effects have been documented, which Thompson suggests require practical actions to address these plastics issues. He outlines proposed areas of intervention – including green chemistry and redesign of products – that might be seen as practices that plastics now provoke or force through their ongoing proliferation and problematic rematerialization of bodies and environments.

New articulations

In the final section of this collection, Part IV, we bring together chapters that consider the new articulations that are emerging or might emerge as speculative practices responding to the material politics of plastics. These are practices that are not necessarily oriented towards ideal solutions, but rather bring us back to the challenge articulated earlier in this introduction to think about more inventive forms of problem making. At what point do plastics become evident as material events that force thinking and spark new types of political engagement? James Marriott and Mika Minio-Paluello (members of Platform, a London-based arts, human rights and environmental justice organization) begin this section with a discussion of the prehistory of plastics in the form of petroleum and its distribution across Azerbaijan to Germany and England. Marriott and Minio-Paluello craft their discussion of the contentious material infrastructures that are essential to the substance and making of plastics by opening a plastic carton of ice cream, and then tracing the geopolitical commitments and invisible violence that have contributed to the material form and availability of this consumer good. Is it possible, these authors ask, to begin to think of consumer goods that are not reliant on oil?

Part of the process of making plastics evident as a matter of concern may involve bringing citizen scientists into the fold of environmental science in order to study the spatial variation and chemical risks associated with plastics in the environment. In his collaborative project, International Pellet Watch, Shige Takada asks volunteers to collect and return by post pellets and plastic fragments that collect on shores across the world. These microplastics are valuable geographic samples because they can be tested for concentrations of POPs. From these widely gathered and mailed-in pellets, International Pellet Watch has generated maps that document the spread and concentration of plastics through seas worldwide. Bringing plastics and practices of studying plastics to wider publics is a process whereby plastics may be seen to be contributing to the emergence of distinct types of scientific practice for studying these matters of concern.

If Takada captures shifts in practices for studying and reporting on plastics, then Jennifer Gabrys proposes that the multiple participants involved in working through and variously breaking down plastics might also begin to be taken into account in the emerging plastic environments of oceans. 'Carbon workers' is a term that she adopts to develop a strategy for making evident the multiple more-than-human entities that are working through leftover plastics that collect in seas, and to describe the practices, processes and politics that emerge when oceans are effectively reconstituted through plastics. In the final chapter of this section, the ways in which plastics are not just external objects of study – epitomizing consumer cultures of disposability – but also material agents that rework bodies and environments, become increasingly evident. What are the practices that sustain our plasticized bodies and environments, and what are the consequences of these entanglements as plastics and plasticizers circulate, break down and transform bodies and environments over time?

In this range of plastic encounters, objects and events, we begin to assemble an account of plastic as a transformative, multiply constituted material that contributes to the emergence of distinct types of practices and political engagements. The plasticity of plastics – as a material in process – emerges and generates distinct responses as captured here, whether through interdisciplinary study, creative practice or proposals for political action. If this edited collection brings one thing to the multiple engagements with plastics that have been generated over the decades since its post-war proliferation, it is to ask how plastics as a material-political force will spark new types of collective engagements with our contemporary and future material worlds.

We would like to thank the contributors to this volume for participating in this ongoing plastics conversation and interdisciplinary experiment. We would also like to thank the Centre for the Study of Invention and Social Process, the Department of Sociology and the Department of Design at Goldsmiths, University of London, and the Centre for Critical and Cultural Studies at the University of Queensland, for funding provided in support of the initial *Accumulation* event in June 2011, and this subsequent publication.

References

Barthes, R. (1972 [1957]) 'Plastic', in *Mythologies*, trans. A. Lavers, New York: Farrar, Straus and Giroux, 97–99

Bennett, J. (2010) *Vibrant Matter*, Durham, NC: Duke University Press.

Braun, B. and Whatmore, S. (eds) (2010) *Political Matter: Technoscience, Democracy, and Public Life*, Minneapolis, MN: University of Minnesota Press.

Carter, P. (2004) *Material Thinking: The Theory and Practice of Creative Research*, Melbourne: Melbourne University Press.

DeLanda, M. (2005) 'Uniformity and Variability: An Essay in the Philosophy of Matter', museum.doorsofperception.com/doors3/transcripts/Delanda.html (accessed 7 March 2012).

Fenichell, S. (1996) *Plastic: The Making of the Synthetic Century*, New York: Harper Collins.

Fraser, M. (2010) 'Facts, Ethics and Event', in C. Bruun Jensen and K. Rödje (eds) *Deleuzian Intersections in Science, Technology and Anthropology*, New York: Berghahn Books, 52–82.

Freinkel, S. (2011) *Plastic: A Toxic Love Story*, New York: Houghton Mifflin Court.

Gabrys, J. (2011) *Digital Rubbish: A Natural History of Electronics*, Ann Arbor, MI: University of Michigan Press.

Hawkins, G. (2010) 'Plastic Materialities', in B. Braun and S. Whatmore (eds) *Political Matter: Technoscience, Democracy and Public Life*, Minneapolis, MN: University of Minnesota Press, 119–38.

Manzini, E. (1986) *The Material of Invention*, Milan: Arcadia Edizioni.

Meikle, J.L. (1995) *American Plastic: A Cultural History*, New Brunswick, NJ: Rutgers University Press.

Michael, M. (2012) '"What are We Busy Doing?" Engaging the Idiot', *Science, Technology & Human Values* 37(5): 528–54.

Stengers, I. (2005) 'Introductory Notes on an Ecology of Practices', *Cultural Studies Review* 11(1): 183–96.

——(2010) 'Including Nonhumans in Political Theory: Opening Pandora's Box?' in B. Braun and S. Whatmore (eds) *Political Matter: Technoscience, Democracy, and Public Life*, Minneapolis, MN: University of Minnesota Press, 3–34.

Thompson, R.C., Swan, S.H., Moore, C.J. and vom Saal, F.S. (2009) 'Our Plastic Age', *Philosophical Transactions of the Royal Society B: Biological Sciences* 364 (1526): 1973–76.

Thrift, N. (2006) 'Space', *Theory, Culture & Society* 23(2–3): 139–55.

Whitehead, A.N. (1929) *Process and Reality: An Essay in Cosmology*, New York: The Free Press.

Part I

Plastic materialities

1 Plastics, materials and dreams of dematerialization

Bernadette Bensaude Vincent

'Plastics happen; that is all we need to know on earth.' This remark is extracted from *Gain*, a novel by the American writer Richard Powers (1998: 395). The novel gives an account of a successful family business that has grown into an international chemical company. A woman, Laura Bodey, who lives nearby the chemical plant, finds she is dying from ovarian cancer, which is presumably induced by substances produced by the company. To her ex-husband, who has advised her to sue the company, Laura replies that, even if the products manufactured by this plant did actually cause her condition, they have given her everything else and moulded her life. It is therefore impossible to balance the costs and gains of plastics. In her view, it does not make sense to blame plastics because they are an integral part of our world, of our lives.

In quoting Laura's reply in *Gain*, Philip Ball (2007: 115) comments that, 'Plastic stands proxy for all our technologies: Plastics generated an entire industrial ecosystem, a technological large-scale-system, which can no longer be controlled'. Taking Ball's stance in a different direction, in this chapter I will argue that plastics have also shaped a new concept of technological design and a specific relation between humans and materials. In particular, they have encouraged the dream of dematerialized and disposable artefacts.

Plastics are more than just ubiquitous manufactured products that are used all over the world. As plastics began to spread in the daily experience of billions of people, new concepts of design were developed that reshaped our view of nature and technology. The phrase 'Plastic Age' – often used to characterize the twentieth century – has been modelled on the epochal categories of Stone Age and Iron Age. Such phrases suggest that the materials used for making artefacts shape civilizations, and that new materials propel a new age. Although our experience of materials is often occulted in daily life by the prevalence of the shapes and functions of the artefacts we use – phones, computers, automotive cars, aircraft – materials *do* matter. They are the core of technological advances and artistic creations; they drive economic exchange and the social distribution of wealth. Each substitution of a material for another one – for instance, iron, aluminium and plastics – engages new relations between nature and artifice, and determines specific relations between science and technology. Cultural historians have described the interaction between plastics and American civilization. For Robert Sklar (1970), the Plastic Age started after World War I when the traditional values of refined society gave way to mass culture, while Jeffrey Meikle (1995) convincingly argues that plastics gradually came to be identified with the American way of life and culture in the second half of the twentieth century, with the emergence of new aesthetics and new societal values.

This chapter aims to provide a better understanding of the interplay between the materiality of plastics and their anthropological dimensions. Previous materials, such as glass, wood and aluminium, are referred to by the name of the stuff of which they are made. By contrast, the common name of synthetic polymers derives from one of their physical properties. The adjective 'plastic' may be a predicate of humans as much as it is of things. The phrase 'Plastic Age' was already in use in the 1920s in the title of a film, and seems to refer to the malleable teenage years, when someone can be changed through life experience. A few years later, in his *Chemistry Triumphant*, William J. Hale announced the 'Silico-Plastic Age' (Hale 1932). The linguistic preference for the term 'plastic' is an indicator that plasticity gained a cultural meaning in the twentieth century. This requires a closer look at the physical and chemical properties of the class of materials gathered under the umbrella 'plastics', as well as at their production process. The entanglement between material, technical and cultural aspects shapes artefacts themselves, and reconfigures the relationship between nature, artefacts and culture.

Following a brief historical sketch about the emergence of plastics-as-plastics and reinforced plastics, the chapter will describe how synthetic polymers contributed to the emergence of a new relationship between technology and matter as they generated the concept of materials by design and 'materials thinking' – a new approach to materials in technological design. The next section looks more closely at the cultural values associated with the mass consumption of plastics, such as lightness, superficiality, versatility and impermanence. I will emphasize the utopian dimension of plastics and the striking contrast between the aspirations to dematerialization or impermanence and the neglected process of material accumulation upstream and downstream, which are respectively the precondition and the consequence of the Plastic Age. Finally, taking up the traditional issue of the relations between the natural and the artificial, I will consider how plastics are reconfiguring the contemporary vision of nature.

Expanding technological capabilities

In the twentieth century, plastics have replaced and displaced wood and metals in many commercial applications. This was by no means a natural and easy movement of substitution. While natural gums and resins such as *gutta percha* were manufactured in the nineteenth century for their insulating properties in electrical appliances, semi-synthetic polymers – such as Parkesine, presented by Alexander Parkes at the London World Exhibition in 1862, and the celluloid manufactured by John Wesley and Isaiah Hyatt in the 1870s – were promoted as alternatives to more conventional solid materials. Lightness and versatility were their most striking novelty. Celluloid was described as a 'chameleon material' that could imitate tortoise-shell, amber, coral, marble, jade, onyx and other natural materials. It could be used for making various things, such as combs, buttons, collars and cuffs, and billiard balls. However, as the historian Robert Friedel (1983) argues, Parkesine and celluloid did not bring about a revolution and did not easily overtake more traditional materials. Celluloid was viewed as just one of myriad 'useful additions to the arts' (Friedel 1983: xvi). Iron, glass and cotton continued to be produced in the millions of tons, while the light celluloid never exceeded hundreds of tons. In addition, the fact that celluloid made out of cellulose and camphor could be given a variety of shapes, colours and uses did not strike consumers as a sign of superiority; on the contrary, its versatile and multipurpose nature was viewed as a major imperfection.

The alliance between one material and one function – still visible in common language when we use phrases such as 'a glass of wine' – was seen as a mark of superiority. This traditional view of nature was reminiscent of Aristotle's view when he claimed that the knives fashioned by the craftsmen of Delphi for many uses were inferior to nature's works because 'she makes each thing for a single use, and every instrument is best made when intended for one and not for many uses' (Aristotle n.d.: 1252b). In this traditional view, multifunctional instruments are for barbarians who don't care for perfection, whereas distinction and discrimination signify the perfection and generosity of nature. Eventually – and despite its flammability – celluloid managed to win a place on the market when it was recognized that it was ideal for a number of applications, such as photographic films. Materials meeting all demands, purposes and tastes were not regarded as dignified. Far from being praised as a quality, plasticity was the hallmark of cheap substitutes, forever doomed to imitate more authentic, natural materials. It is only in retrospect, in view of the ways of life and the values generated during the Plastic Age, that we have come to value multifunctional artefacts.

Today, plastics are no longer considered cheap substitutes. They are praised because they can be moulded easily into a large variety of forms and remain relatively stable in their manufactured form. Certainly, the success of plastics-as-plastics is due to the active campaigns of marketing conducted by publicists who promoted them as materials of 'protean adaptability' that could meet all demands and bring comfort and luxury into everyone's reach (Meikle 1995). Chemical companies in America presented plastics as a driving force towards the democratization of material goods. In the 1930s, chemical substitutes were also praised as pillars of social stability because they provided jobs and fed the market economy: 'a plastic a day keeps depression away' (Meikle 1995: 106).

Enhancing the performances of plastics

In addition to the social benefits expected from plastics, a number of technical aspects related to their process of production account for plastics overtaking more traditional materials. Wood and metals pre-exist the action of shaping them: wood is carved or sculpted; metals are ductile and malleable – they melt at high temperatures, then the molten metal can be cast in a mould or stamped in a press to form components into the desired size and shape. By contrast, plastics are synthesized and shaped simultaneously. The process of polymerization is initiated by bringing the raw materials together and heating them – it is not separate from moulding. In more philosophical terms, matter and form are generated in one single gesture. This specific process is due to the ability of carbon atoms to form covalent bonds with other carbon atoms or with different atoms. Thus, a chain of more than 100 carbon atoms can make a single macromolecule. The resulting thermosetting polymers are rigid, with remarkable mechanical properties; furthermore, unlike celluloid they are not heat sensitive. They are lightweight, have a high strength-to-weight ratio, are corrosion resistant, remain bio-inert, and have high thermal and electrical insulation properties. However, they cannot be reheated and moulded again. Soon, a newer category of polymers came on to the market: these form weaker chemical bonds, and consequently can be reheated, melted and reshaped. These thermoplastic polymers, such as the polyethylene manufactured in the 1930s, are less rigid and more plastic than thermosetting polymers.

The synthetic polymers manufactured after World War II were already more plastic than early plastics and thermoplastics – such as polyethylene, polypropylene, polyester and polyvinyl chloride (PVC) – and undoubtedly had a wide spectrum of applications. However, the plasticity of plastics can still be enhanced because various ingredients are added to the raw materials and included in the process of polymerization. Pigments were regularly added to produce a variety of colours, which became a distinctive feature of plastic materials in the 1930s. Inorganic fillers of silica were also used to make cheaper materials. Other additives can improve various properties: thermal or UV (ultraviolet) stabilizers increase resistance to heat and light; plasticizers are added to make them more pliable or flexible (Andrady and Neal 2009); improved mechanical properties are obtained thanks to the addition of reinforcing fibres. Glass fibres were first added to reinforce plastics in the 1940s for military applications such as boats, aircraft and land mines (Mossman and Morris 1994). Reinforced plastics enabled expansion of the market in plastics in the 1950s for civil applications such as electric insulators and tankers. Initially, reinforced plastics were introduced for the purpose of weight saving and cost reduction in transport and handling. However, they generated a deep change in design, and facilitated a new approach to materials research.

Composites and materials by design

Because the mechanical properties of heterogeneous structures depend upon the quality of interface between the fibre and the polymer, it was crucial to develop additive substances favouring chemical bonds between glass and resin. The study of interfaces and surfaces consequently became a prime concern, and gradually reinforced plastics gave way to the general concept of composite material (Bensaude Vincent 1998). Although most commercial composites are made of a polymer matrix and a reinforcing fibre, composites may be made of metal and fibre. The concept of the composite that came out of plastics technology has been extended to all materials associating two phases in their structure where each one assumes a specific function: steel or iron is used as a support for toughness; plastics are useful for weight saving; and ceramics are included for heat resistance and stiffness. Creating a composite material means combining various properties that are mutually exclusive into one single structure. Composites were created initially in the 1960s for aerospace and military applications. In contrast to conventional materials with standard specifications and universal applications, they were developed with both the functional demands and the services expected from the manufactured products in mind. Such high-tech composite materials, designed for a specific task in a specific environment, are so unique that their status becomes more like that of artistic creations than standard commodities.

While reinforced plastics were aimed basically at adding the properties of glass fibre or higher-modulus carbon fibres to the plasticity of the polymer matrix, composites did reveal new possibilities and generated innovations. For instance, the substitution of old chrome-steel bumpers of the cars of the 1950s for plastic bumpers did not immediately entail the cost reduction that was expected because the composite had opened new avenues for change. Manufacturing and shaping the chrome steel were two successive operations; in the case of plastic they became one and the same process. Car designers were consequently free to curve the bumper along the line of the shell. Instead of a separate part that had to be manufactured independently and then welded to the car, the shell was integrated with the body of the car like a protective second skin. In addition to protection, other functions could similarly be integrated. Thus, ventilators and radiator grilles were combined with the same unit at the front. Integration proved useful because it reduced the number of parts and assembly steps. New concepts thus emerged that gradually integrated more and more functions into the same structural part. However, local change in the material structure of one part called for redesigning the whole automotive structure and, thanks to the synergy between structure, process and function, composites contributed to the development of a new specific approach to designing materials. The interaction of the four variables – structure, properties, performances and processes – is such that changes made in any of the four parameters can have a significant effect on the performance of the whole system and require a rethinking of the whole device. Engineers had to give up the traditional linear approach to innovation ('given a set of functions, let's find the properties required and then design the structure combining them'), and convert to 'materials thinking'. They simultaneously had to envision structure, properties, performance and process.

Thanks to the enhancement of the intrinsic chemical and physical properties of plastics through materials thinking, their market expanded to profitable and successful applications in transportation, sports items and a wide range of other products. Materials thinking also played a crucial part in the emergence of a new relationship with materials and matter in general. For materials designers, 'materials thinking' basically refers to a systems approach – a new method of design that takes into account all parameters simultaneously rather than sequentially. It has no connection with the phrase 'material thinking' in the vocabulary of social scientists, which mainly refers to the materiality of thinking (Carter 2004; Thrift 2006). Despite the divergence of references, the rapprochement between the two contexts is interesting in terms of opening the question of the meaning to be given to this new practice of design. Social scientists use the expression 'material thinking' in order to emphasize the active participation of materials in the mental activity of thinking. Similarly, the designers of artefacts could insist on the role of the physical and chemical properties of plastics that afford new opportunities in terms of design. They could emphasize that materials become active participants in the design process rather than passive objects of manipulation. However, in their discourse, materials have no say in the creative process. On the contrary, engineers and designers seem to emphasize that materials are no longer a prerequisite for design, as they adopt the phrase 'materials by design'. This phrase suggests that they are emancipated from the constraints and resistance of matter.

Materials themselves can be purposely tailored to perform specific tasks in specific conditions. For instance, in the 1960s, space rockets required never-seen-before combinations of properties: they had to be lightweight and resistant to both high temperatures and corrosion. Early composite materials were designed for such applications, and a number of them have been transferred successfully to everyday commodities such as sports articles or clothes. Materials are no longer a prerequisite for the design of artefacts, and would no longer limit our possibilities of creation. Thanks to the enhanced plasticity of composites, designers could feel emancipated from the constraints of matter, free to create artefacts, buildings or haute couture clothes according to their own inspiration.

Composites encouraged the quest for the ideal material, with a structure in which each component would perform a specific task according to the designer's project. Matter came to be presented as a malleable and docile partner of creation – a kind of Play-Doh in the hands of the clever designer who informs matter with intelligence and intentionality. Just like the *demiurgos* in Plato's *Timaeus*, the material engineer can impose forms on a passive, malleable *chora*. For instance, in the 1990s, a French company manufacturing sheet-moulding compounds for making composites advertised its products with the image of a plastic toy car and the following comment: 'What is fantastic with Menzolit play doughs [sic] is that one can press, inject, twist them, they lend themselves to all your ideas.' The plastic resin being shaped and informed by human intelligence becomes a smart composite material. The ad proudly concluded: 'Grey matter [is] the raw material of composite materials' (Menzolit 1995).

Designing materials with built-in intelligence is the ultimate goal of a number of research programmes launched in the 1990s. Smart or intelligent materials are structures with properties that can vary according to changes in their environment. They are plastic insofar as they can adjust to changing conditions or self-repair in case of damage. For example, materials with a chemical composition that varies according to their surroundings are used in medicine to make prostheses. This requires them to have embedded sensors (for strain, temperature or light) and actuators so that the structure becomes responsive to external stimuli.

The stuff that dreams are made of

Plasticity, the distinctive property of synthetic polymers, has permeated through culture. The French philosopher Roland Barthes (1971) devoted a few pages to plastics in his review of the mythologies of modernity. 'Plastics', he wrote, 'are like a wonderful molecule indefinitely changing' (Barthes 1971: 171–72). Plastics are shapeless; they have pure potential for change and movement. They connote the magic of indefinite metamorphoses to such a degree that they lose their substance, their materiality, to become virtual reality. Plastics have thus encouraged the utopia of an economy of abundance that could consume less and less matter by using cheap, light, high-tech plastics. Although Barthes witnessed only the debut of the flood of cheap fashionable and disposable products especially designed to become obsolete after a few uses, he saw the coming of a new relation of our culture to time. Whereas gold or diamond conveys a view of permanency and eternal faith, plastics epitomize the ephemeral, the ever changing. They invite us to experience the instant for itself as detached from the flux of time.

In his remarkable study of plastics in American culture, Jeffrey Meikle (1995) emphasizes that plastics have often been presented as 'utopian materials', and that they gradually came to epitomize a kind of dream world. Such a utopian world is played out not only in the rapprochement between plastics and Disney World, which relies on the abundance of fibreglass-reinforced polyester structures in the amusement park at Orlando, but also through the material-cultural values developed along with the use of everyday plastic objects, from BIC pens to razors, telephones and credit cards. In this way, the daily experience of plastics transformed American culture: 'Increasingly that culture was seen as one of plasticity, of mobility, of change, and of open possibility for people of every economic class' (Meikle 1995: 45). Indeed, the counter-culture movement, which criticized the American way of life, used the term 'plastic' as a metaphor for superficial and inauthentic people whose lives were driven by a passion for consumption and change.

Such critics could also point to the inherent paradox of plastics. These light, colourful and cheap materials, apparently liberated from the constraints of gravity, from rigid shapes and duration, are inextricably linked to the accumulation of huge quantities of matter and energy. As Jean Baudrillard (2000) points out, plastics instantiate the contradictions of a society oriented towards the mass manufacture of more and more disposable products. About 300 million tons of plastics are produced each year. These ephemeral commodities generate tons of durable waste, since thermoplastics can persist for extended periods of time in the environment (Barnes *et al.* 2009). From urban suburbs to the most remote places in the countryside, they have invaded the natural habitats of living species on earth and in the oceans. Furthermore, as most synthetic polymers are made out of fossil fuels, they use about 4 per cent of the world's oil in material substance (and 4 per cent of the world's oil in the form of energy for manufacturing).

Plastics irreversibly consume the vestiges of plants accumulated over thousands of years. The two processes of accumulation surrounding the short life of plastic commodities clearly indicate that their ephemeral character is delusory. Despite its hedonistic inclinations, the Plastic Age developed a mathematical notion of time as an abstract space consisting of a juxtaposition of discrete points or instants, blurring all issues of persistence and permanence. Plastics are supposed to be ephemeral only because – like the flying arrow of Zeno's paradox commented on by Bergson (1946) – they are supposed to be at rest, as moments of being. By contrast, our Plastic Age confronts the issue of duration. The ephemeral present of plastics is not just an instant detached from the past and the future. It is the tip of a heap of memory, the upper layer of many layers of the past that have resulted in crude oil stored in the depths of the soil and the sea. The cult of impermanence and change has been built on a deliberate blindness regarding the continuity between the past and the future. Plastics really belong to Bergson's (1946) duration; they cannot be abstracted from the heterogeneous and irreversible flux of becoming. The present is conditioned by the accumulated traces of the past, and the future of the earth will bear the marks of our present. While the manufacture of plastics destroys the archives of life on the earth, its waste will constitute the archives of the twentieth century and beyond.

Plastic nature

According to cultural historians, the Plastic Age culminated with the fashion for artificial fabrics, paintings and dyes. In the plastic items manufactured in the 1960s and 1970s, shining, fluorescent and flashy surfaces prevailed over the traditional preference for pastel colours that looked more natural or genuine. The cult of the artificial exemplified by Andy Warhol paintings broke with the early plastics, which desperately attempted to imitate wood, horn, shell or ivory in appearance and colour. They had no intrinsic value – they were praised only for their cheapness and their potential for the democratization of comfort. They were also occasionally valued because synthetic substitutes could spare the life of tortoises, elephants and baby seals. For instance, Williams Haynes (1936: 155) claimed that 'The use of chemical substitutes releases land or some natural raw material for other more appropriate or necessary employment'. The synthetic was thus a useful detour in the conservation and protection of nature.

The Plastic Age radically transmuted the cultural values attached to the natural and the artificial, and reinforced the cultural stereotype associating chemists with Faust or the alchemists who challenged nature. At first glance, it could be expected that, by design, the light, quasi-immaterial materials would reinforce the culture of the artificial initiated by thermoplastics in the mid-twentieth century. What could be more unnatural than composite materials as light as plastic with the toughness of steel and the stiffness or heat resistance of ceramics? Like the centaurs invented by the Ancients, they combined different species into one body, into their inner structure. They could consequently revive the mythical figures of Prometheus or Faust. Indeed, the Promethean view of engineers 'shaping the world atom by atom' has been revitalized by the promoters of nanotechnology. The slogan of the US 2000 National NanoInitiative announced an era when materials would be designed and engineered bottom-up, with each part of the structure performing a specific task (Bensaude Vincent 2010). The ambition to overtake nature with our artefacts is still very much alive today.

It is nevertheless counterbalanced by a back-to-nature movement that emerged in the 1980s. The more pressing the quest for high-performance and multifunctional plastics, the more materials chemists and engineers turned to nature for inspiration. Most of the 'virtues' embedded in materials by design – such as minimal weight, multifunctionality, adaptability and self-repair already exist in natural materials. Amazing combinations of properties and adaptive structures can be found in modest creatures such as insects and spiders. Spider webs attracted the attention of materials engineers because the spider silk is made of an extremely thin and robust fibre, which offers an outstanding strength-to-weight ratio. Wood, bone and tendon have a complex hierarchy of structures, with each different size scale – from the angstrom to the nanometre and micron – presenting different structural features. Their remarkable properties and multiple functions are the result of complex arrangements at different levels, where each level controls the next one. Nature displays a level of complexity far beyond any of the complex composite structures that materials scientists have been able to design. In addition, nature designs responsive, self-healing structures that quickly adapt to changing environments. Above all, the plastic structures designed by nature avoid the vexing issue raised by human-made plastics, namely accumulating tons of litter all around the world. They are degradable and recyclable.

Finally, what materials designers most envy is nature's building processes. Synthetic chemists managed to get polymerization and moulding, matter and form, into one single operation. Nature goes even further, thanks to the self-assembly of molecules. While synthetic polymers are built with strong covalent bonds, molecular self-assembly is a spontaneous organization of molecules into ordered and relatively stable arrangements through weak non-covalent interactions. Molecular self-assembly is extremely advantageous from a technological point of view, because it generates little or no waste and has a wide domain of application (Whitesides and Boncheva 2002). Self-assembly appears to be the holy grail for designing at the nanoscale, where human hands and conventional tools are useless. It is the key to a new age: 'The Designed Materials Age requires new knowledge to build advanced materials. One of the approaches is through molecular self-assembly' (Zhang 2002: 321).

Because molecular self-assembly is ubiquitous in nature, nature seems to capture all the attributes of plastics. Whereas in the early twentieth century natural structures were characterized as rigid, stiff, resistant and resilient in contrast to synthetic polymers, one century later the same natural structures investigated at the nanoscale are characterized as 'soft machines' (Jones 2004): highly flexible, adaptive, complex and ever changing.

Despite their admiration for nature's achievement, biomimetic chemists are not inclined to revive natural theology and its celebration of 'the wonders of nature'. Rather, biomimicry proceeds from a technological perspective on nature. Nature is depicted as an 'insuperable engineer' that took billions of years to design smart materials. They study the structure of biomaterials and the natural process of self-assembly with the conviction that nature has worked out a set of solutions to engineering problems. With its exquisite plasticity, nature affords a toolbox to inventive designers of advanced materials. Atoms and molecules are functional units useful for making nano-devices such as molecular rotors, motors or switches. Biopolymers provide smart tools: the two strands of DNA are used to self-assemble nano-objects; liposomes are used as drug-carriers. Living organisms such as bacteria are being re-engineered or even synthesized to perform technological tasks. 'E-coli moves into the plastic-age' was the title of one research news item announcing that plastics that are part of our lifestyle would be synthesized by E-coli bacteria with no waste disposal, and no more pollution or contamination of the environment (Lee 1997).

Conclusion

In following the migrations of the term 'plastic' from the realm of materials to the realm of humans and to nature throughout the twentieth century, this chapter has emphasized the interplay between materials and culture. From a view of nature as a stable, rigid order, our culture has shifted to a view of nature as plastic, versatile and based on the ever-changing arrangements of molecular agencies. The success story of plastics, which combined the specific features of synthetic polymers and the markets in which they flourished, deeply reconfigured consumer practices as well as those of design. Because plastics are objects of design, they are more than polymers. The classical terminology of polyethylene, polystyrene, polypropylene, phenol-formaldehyde and so on is not really adequate, since the properties and uses of plastics depend on plasticizers, fillers, UV protectors and the like. The traditional classifications of materials become obsolete when plasticity is so highly praised that design embraces materials themselves. Thus, plastics renewed the ambition of shaping the world according to our purposes with no resistance from nature.

This chapter has also pointed to the blind spots generated by the Plastic Age. In cultivating plasticity as a chief value, the twentieth century had to develop a sort of blindness about the impacts of material consumption on the environment and on the future. Indeed, mass consumption in general requires no concern with the afterlife of commodities, however much the cult of disposability and ephemerality associated with plastic reinforced and perpetuated this denial. The cultural history of plastics must be completed by agnotology studies pointing to the social construction of ignorance necessary for the mass diffusion of plastics (Proctor and Schiebinger 2008). This sort of ignorance is a denial – a self-deception – that allows us to live in a fool's paradise.

Although the twenty-first century seems to be more aware of environmental issues and more concerned with the future, plastics retain their utopian nature. Plastic items may have acquired a very bad reputation for many people, but the concept of plastic as malleable matter is still extremely attractive. The emerging economy of biopolymers and biofuels designed at the molecular level is based on the vision of nature as a limitless field of potentials. Design from bottom up, proceeding from the ultimate building blocks of nature, is supposed to meet no resistance and to afford a free space for creativity. It encourages the view of matter as purely plastic, passive and docile, subject to the designer's purposes. The techno-utopia of the Plastic Age is not over. It continues through the denial of the constraints imposed by matter and nature's laws. Just as 'the light dove, cleaving the air in her free flight, and feeling its resistance, might imagine that its flight would be easier in empty space' (Kant 1965: 48), contemporary designers cherish Plato's illusion that we could be free from matter and venture beyond it on the wings of ideas. In paraphrasing Kant's (1965) criticism of Plato, one could say that the Plastic Age will be over when the dove-designer realizes that resistance might serve as a support upon which to take a stand and to which he could apply his powers.

Acknowledgements

This paper benefited from the support of the French-German programme 'Genesis and Ontology of Technoscientific Objects' (ANR09-FASHS-036-01).

I am indebted to Dr Robert Bud for the references on early occurrences of 'Plastic Age'.

References

Andrady, A.L. and Neal, M.A. (2009) 'Applications and Social Benefits of Plastics', *Philosophical Transactions of the Royal Society B* 364: 1977–84.

Aristotle (n.d.) *Politics*, I, 2, 1252b, www.perseus.tufts.edu/hopper/text?doc=Perseus%3Atext%3A1999.01.0058%3Abook%3D1%3Asection%3D1252b (accessed 8 October 2012).

Ball, P. (2007) 'Chemistry and Power in Recent American Fiction', in J. Schummer, B. Bensaude Vincent and B. Van Tiggelen (eds) *The Public Image of Chemistry*, London: World Scientific, 97–122.

Barnes, D.K., Galgani, F., Thompson, R.C. and Barlas, M. (2009) 'Accumulation and Fragmentation of Plastic Debris in Global Environments', *Philosophical Transactions of the Royal Society B* 364: 1985–98.

Barthes, R. (1971 [1957]) *Mythologies*, 2nd edn, Paris: Seuil.

Baudrillard, J. (2000 [1968]) *Le système des objets*, 2nd edn, Paris: Gallimard.

Bensaude Vincent, B. (1998) *Éloge du mixte: Matériaux nouveaux et philosophie ancienne*, Paris: Hachette Littératures.

——(2010) 'Materials as Machines', in A. Nordmann and M. Carrier (eds) *Science in the Context of Application*, Dordrecht: Springer, 101–14.

Bergson, H. (1946) *The Creative Mind: An Introduction to Metaphysics*, New York: Kensington.

Carter, P. (2004) *Material Thinking: The Theory and Practice of Creative Research*, Melbourne: Melbourne University Press.

Friedel, R. (1983) *Pioneer Plastic: The Making and Setting of Celluloid*, Madison, WI: University of Wisconsin Press.

Hale, W.J. (1932) *Chemistry Triumphant: The Rise and Reign of Chemistry in a Chemical World*, Baltimore, MD: Williams & Wilkins in cooperation with The Century of Progress Exhibition.

Haynes, W. (1936) *Men, Money and Molecules*, New York: Doubleday, Doran & Co.

Jones, R. (2004) *Soft Machines*, Oxford: Oxford University Press.

Kant, I. (1965 [1781]) *The Critique of Pure Reason*, trans. Norman Kemp Smith, New York: St Martin's Press.

Lee, S.Y. (1997) 'E-coli Moves into the Plastic Age', *Nature Biotechnology* 15(1): 17–18.

Meikle, J.L. (1995) *American Plastic: A Cultural History*, New Brunswick, NJ: Rutgers University Press.

——(1996) 'Beyond Plastics: Postmodernity and the Culture of Synthesis', in Gerhard Hoffmann and Alfred Hornung (eds) *Ethics and Aesthetics: The Moral Turn of Postmodernism*, Heidelberg: C. Winter, 325–42.

——(1997) 'Material Doubts: The Consequences of Plastic', *Environmental History* 2 (3): 278–300.

Menzolit (1995) Advertisement in *Composites, plastiques renforcés, fibres de verre textile* 8 (March–April): 3.

Mossman, S. and Morris, P. (eds) (1994) *The Development of Plastics*, London: The Science Museum.

Powers, R. (1998) *Gain*, New York: Picador.

Proctor, R.N. and Schiebinger, L. (eds) (2008) *Agnotology: The Making and Unmaking of Ignorance*, Stanford, CA: Stanford University Press.

Sklar, R. (ed.) (1970) *Plastic Age (1917–1930)*, New York: George Braziller.

Thompson, R.C., Swan, S.H., Moore, C.J. and vom Saal, F.S. (2009) 'Our Plastic Age', *Philosophical Transactions of the Royal Society B: Biological Sciences* 364 (1526): 1973–76.

Thrift, N. (2006) 'Space', *Theory, Culture & Society* 23(2–3): 139–55.

Whitesides, G.M. and Boncheva, M. (2002) 'Beyond Molecules: Self-assembly of Mesoscopic and Macroscopic Components', *Proceedings of the National Academy of Science* 99: 4769–74.

Zhang, S. (2002) 'Emerging Biomaterials through Molecular Self-assembly', *Biotechnology Advances* 20: 321–39.

2 Process and plasticity

Printing, prototyping and the prospects of plastic

Mike Michael

Introduction

This chapter[1] is concerned with the emerging future of plastic as it enters into the home in a new guise. With the rise of new technologies and their co-constitutive discourses, plastic is being opened up – 'democratized' – as a newly manipulable material. At the same time, this 'democratization' requires particular technical skills of production and consumption. As such, the chapter is interested in a specific material – plastic – politics in which the capacities of humans and non-humans together are reconfigured, enacted and performed in particular ways (e.g. Braun and Whatmore 2010).

Now, obviously, plastic is a long-standing cohabitant in most Western homes: it has become a stock material out of which a plethora of products are constructed, or partly constructed. A quick survey of the products in David Hillman and David Gibbs's (1998) *Century Makers: One Hundred Clever Things We Take for Granted Which Have Changed Our Lives over the Last One Hundred Years* reveals just how many have plastic as an integral component: hairdryers, toasters, washing machines, irons, frozen food, ballpoint pens, training shoes, Velcro®, child-resistant caps, LCDs. Yet these are very much *products*. They are consumables that have emerged from design houses and factory production lines. These are not easily made at home: even the smaller (plastic) component parts that go to make up the final product cannot be easily, if at all, manufactured within the domestic – or craft – sphere, as the designer Thomas Thwaites's (2011) Toaster Project, during which he attempted to construct a toaster from scratch, demonstrates only too clearly.

There is no doubt an element of stating the obvious in the foregoing, and yet it would appear that this obviousness is in the process of being overturned. In the specific instance of plastic, this seeming shift is, it can be argued, a partial outcome of the emergence of rapid prototyping or 3D printing technology that uses plastic as one of its primary materials. More specifically, the 3D printer is moving out of the domain of professionals and specialists (designers, model makers, product developers, manufacturers) and into the space of the home. Or rather, this movement is beginning to reconfigure such spaces – it is a movement that, potentially, peculiarly and problematically re-spatializes the domestic sphere and the workplace (where workplace throughout this chapter refers to industrial or manufacturing rather than, for example, office settings). Moreover, it is, potentially at least, re-articulating other 'global' inter-relations: to domesticate production is also to have an impact on contemporary global patterns of manufacturing (from China to the home) and on global ecological patterns of waste production (from the irredeemably broken to the readily reparable). The emergence of the domestic 3D printer is thus instrumental in generating a series of complex, sometimes ironic, relationalities that range across, for instance, craft and expertise, informed materials and 'disinformed' humans, consumption and production, global and local, sustainability and profligacy, expectation and fantasy.

The implications of 3D printing are clearly enormous and extend well beyond the scope of this chapter, which restricts itself mainly to a discussion on the potential impacts on the domestic user. So, in what follows, there is an initial consideration of the apparent sociotechnical position of plastic in contemporary Western societies. This leads into a discussion of the complex relations of plastic to 'plasticity', not least as these inflect with issues of spatial divisions between workplace and home, and the role of craft and skill within such divisions. In the subsequent section, the theoretical underpinnings of the chapter will be explicated and the debts to such writers as Whitehead and Deleuze laid bare. There will then be a discussion of 3D printing, especially with regard to the ways plastic has been 'eventuated' in relation to various contested futures and the contrasting capabilities of both machines and humans. In the next section, we go on to address the apparent rise of the domestic 3D printer. In light of the accompanying claims made for its utility, plastic takes on a role in the renewed blurring of the boundaries between home and workplace, production and consumption. At the same time, new boundaries look set to emerge. In the final section, the meaning of this patterning of de- and re-territorialization is further interrogated, not least in the light of other possible futures of 3D plastic printing.

So little plasticity in plastic?

Without retracing a history of plastic, we can say that plastic is a material that is quintessentially industrial. From the extraction of the source material (e.g. oil), through the chemical process of its production, and onto the procedures of design and manufacture by which plastic artefacts are made, plastic belongs to the realm of the factory. It is not a material that is easily manipulable beyond the specialist combinations of machines and humans that chemically compose and process, dye, extrude, mould and finish plastic goods. Plastic, as an icon of post-war Fordist industrialization, and even the batch production of post-Fordism, has a role in the policing of the spatial boundaries between home and factory (Lefebvre 1974).

This remoteness of plastic production, despite the many intimacies of its consumption, can be reframed in terms of the idea of craft. In the first of a series of videos that accompanied the Victoria & Albert Museum's exhibition the 'Power of Making', there is a comment by a flutemaker: 'I think craft or making things with your hands is fundamental to being human.' Of course, this can be read in a number of ways: an insistence on the dignity of a particular sort of labour; a lament over the rise and predominance of industrial or Fordist production; a definition of the human in terms of a practical – handed – engagement with the world (see Sennett 2008). Yet this all assumes that the appropriate materials and tools are to hand. Or rather, it assumes a special sort of relationality between maker, tool and material. Some materials do not lend themselves, or lend themselves in very limited ways, to the potential craftsperson. These materials can be said to be 'informed' (Barry 2005; Bensaude Vincent and Stengers 1996) insofar as, paraphrasing Barry (i.e. replacing 'pharmaceuticals' with 'plastics'):

> [Plastics] companies do not produce bare molecules … isolated from their environments. Rather, they produce a multitude of informed molecules, including multiple informational and material forms of the same molecule. [Plastics] companies do not just sell information, nor do they just sell material objects … The molecules produced by a [plastics] company are already part of a rich informational material environment, even before they are consumed.
>
> (Barry 2005: 59)

In the case of plastic, this informedness partially manifests itself through exclusion: only certain industrial actors have the capacities to make and mould plastic. The (domestic) human hand is marginalized. As hinted at above, plastic is perhaps the example *par excellence* of what, from the perspective of the domestic and of craft, is an abject relationality – the impoverished possibilities of re-inventing and re-informing plastic. In sum, plastic is a material with a 'composition' (where 'composition', read through a Whiteheadian lens, implies that the chemical properties of plastic are mediated by the co-presence of a nexus of technologies, systems, skills, environments and so on) that precludes manipulation outside of an industrial setting. Thus, while plastic affords innumerable uses in its forms, it imposes considerable constraints in its substance (see Shove *et al.* 2007).

However, we need to tread warily here. While plastic does indeed take on many discrete forms and functions, these can be domestically adapted to alternative uses. Examples abound. At one end of the spectrum there are basic reuses: carrier bags become mini-binbags; margarine tubs, yoghurt cartons, sawn-off plastic bottles become containers for screws, for nuts and bolts, for soil and seeds and so on. At another end of the scale, plastic cutlery is transformed into art, and the square cup-shaped piece of plastic that once attached a D-lock to a bike frame is turned into an elegant and efficient attachment-cum-receptacle for the supporting arms of a resurrected child's bicycle seat (though I say it myself). However, these are gross interventions: pieces, or existing shapes, of plastic are redeployed, and at most are cut or sawn or melted down to size (or glued or taped together up to size). There is no constitutive reshaping or recasting, say, of the many different bike light plastic fitments; once these break, one can't mend them, or manipulate other fitments to accommodate now fitment-less bicycle lights. The upshot is a collection of bicycle lights that either remain unused or require ad hoc forms of attachment (such as strapping to handlebars with electrical tape).

In a word, there is little plasticity in plastic, especially if we take plasticity to connote the potential for new or renewed connections to be rendered domestically (i.e. outside of a professional or industrial setting) and thus for the functions of plastic to be recovered or altered or adapted or invented. This differs somewhat from, but also supplements, Bensaude Vincent's (2007) formulation of plasticity, which, within a discussion of the changing borders between the natural and the artificial, places its emphasis on the mass production of polymers and the rise of the Plastic Age with its profusion of plastic goods. Accordingly, 'plasticity' signifies the multiplicity of goods and functions that a single material (or, rather, class of materials) can yield. By contrast, nature was marked by inflexibility and limitation.

In the present analysis, the notion of plasticity is partly developed with reference to localization in neurophysiology, where function is mapped onto particular locations on or within the brain such that if a particular area is damaged or destroyed then there is a permanent loss of the correlated function. If the part of the cortex responsible for a specific set of movements is destroyed through stroke, say, the capacity for that movement is lost. By contrast, plasticity allows for neural adaptation in which new connections can be made that 'correct' for the damaged or destroyed area and enable lost functions to be, more or less, recovered. In the case of motor-neurone damage, it has been found that adjacent undamaged areas take over the work of the damaged area.

However, we should take heed of Susan Leigh Star's (1989) observation that the comparative privilege enjoyed by theories of plasticity versus locationalism reflect and mediate the particular sociomaterial conditions of the time (e.g. World War I renders locationalism more 'useful' as a way of dealing – coping – with, and treating, the enormous number of brain-damaged soldiers returning from the front; arguably modern neuroimaging techniques such as fMRI and PET are instrumental in the apparent contemporary resurgence of locationalism). The general point here is that plasticity is itself a plastic concept, its content and utility varying under different circumstances. The present aim is not to map or typify the versions of plasticity but merely to trace some of the specific ways in which the plasticity of plastic might be undergoing change through an emergent nexus of sociomaterial relationalities associated with 3D printing.

Above I argued that we needed to be circumspect about the plasticity of plastic as we move from form to substance, or from the production of variety to the consumption of specific plastic artefacts. However, a note of caution now needs to be sounded. Drawing on a different literature, we can say that the use of plastic tends toward 'standardization'. Its function is scripted (Akrich 1992; Akrich and Latour 1992) into the objects of which it is a part. There is a specified and necessary range of capacities and skills 'built into' (the functioning of) the plastic child-proof bottle cap. Of course, as Latour (1992) has pointed out, these scripts also serve to discriminate against certain bodies – sometimes these discriminations are positive (as in the case of children, obviously enough), sometimes they are negative (as is the case for elderly people with less strength or mobility in their hands). Further, though, such mundane technologies can also invite a certain sort of craft: for instance, they can be used to precipitate resistance, subversion or protest, that is to engender sociomaterial innovation (Michael 2000a).

In addition, there is also the matter of aesthetics (or the aesthetics of matter) to take into account: there is a craft to 'how' one opens a child-proof cap that is partially captured by such terms as elegance, skilfulness, style. This simply points – as much recent literature on consumption does (e.g. Lury 1996) – to the performativity of engagement with objects, plastic or not. Use is as much about expression as utility, and as such it performs not only practical functions but also social relations (often indissolubly so). This performativity is evidenced even in those plastic technologies that seemingly require no skill at all. Michael (2006; also see Halewood and Michael 2008) has suggested that Velcro® is emblematic of unproblematic functionality: as it says on the Velcro® website in relation to Velcro® packages, these are 'designed and engineered for all ages and motor skill levels'.[2] Yet problems abound. Indeed, to make Velcro® work seamlessly, or skill-lessly, a lot of craft has sometimes to be mobilized. Michael recounts the frequent episodes with his daughter when her fine hair would get caught in the Velcro® strips that were attached to her cycling helmet. Over time, and with considerable negotiation, movements became mutually choreographed so that, with some craft, hair and Velcro® remained disconnected. This was a delicate, shared reconfiguration of comportments – it was a collective sociomaterial performance (or instance of heterogeneous performativity) in which not only function but also, indissolubly, an emerging relationship between daughter and father was enacted (also see Mol 2002).

In this section, we have considered how plastic can be portrayed as a material that has served in the differentiation of the sociomaterial spaces of industrial workplace and home, and between the practices of production, craft and consumption. At the same time, we have also noted that these divisions are not simple: that craft attaches to the way that even the most mundane and putatively skill-less plastic artefacts are put to work, not least when craft is understood in terms of a heterogeneous performativity that straddles both the practical and the expressive. As such, we have moved from a view of plastic that, within the confines of the everyday, has a diminished plasticity, to a vision of plastic as occasioning numerous 'small' relationalities that make up a sort of subterranean plasticity in which minute, often unnoticed or unremarked, adaptations and innovations are instituted as a matter of routine. Ironically, these little plasticities serve in the reproduction of plastic's lack of plasticity (see Bowker and Star 1999) – that is to say, they reinforce the impression of plastic's limited plasticity in everyday life. In the next section, we consider how we might better theorize this dynamic, especially in light of the emergence of rapid prototyping or 3D printer technology that can be used to make plastic objects.

From plasticity to plastic event

In the previous section, an attempt was made to understand plastic in terms of its reflection and mediation of particular spatializations (domestic/industrial) and comportments (craft-ful/skill-less). The outcome was a complication of these categories: craft was found in skill-lessness, and consumption of the domestic sphere inflected (to some extent) with the production of the industrial (on this score, also see Cowan 1987). In part, this complication arises because the discussion has been premised on the assumption that there is such a thing as 'plastic' per se. The proliferation of counter-examples and contingencies above indicates that plastic, like all entities, is perhaps more fruitfully regarded in terms of process: it is something that emerges in events – it is eventuated. This formulation derives from the process philosophies of A.N. Whitehead and Gilles Deleuze. Without entering into details, we can propose that what plastic is – its ontology – rests on the sorts of events (actual occasions) of which it is a part and out of which it emerges, and thus on the various social and material elements (prehensions) that come together and combine (concresce) within that event (Whitehead 1929; also Halewood 2011). As such, 'plastic' is always eventuated in its specificity: in other words, there is no abstracted plastic per se to which qualities such as cold, or green, or flimsy, or industrial are attached. Rather, there is flimsy plastic, or green plastic and so on; and any abstracted plastic is itself abstracted in its specificity, say by a chemist or an historian or a designer.

This schema allows us to move away from the abstraction of plastic that was deployed above, and to focus on its concrete eventuations. However, we do need to unpack a little further the notion of event that is being used here. The entities within the event are not simply 'being with' each other, they are also in a process of 'becoming together' (see Fraser 2010) – rather than interacting, they intra-act (Barad 2007). Put another way, there is always an element of uncertainty or openness about the event as the elements become together – what the event 'is' is immanent, it is open to the virtual, subject to de-territorialization – at least in principle (see, for instance, Massumi 2002; DeLanda 2002; Bennett 2010).

Or rather, there are parallel processes of de- and re-territorialization: events simultaneously 'open up' and 'close down'. As Deleuze and Guattari (1988: 10) frame it, 'How could movements of deterritorialization and processes of reterritorialization not be relative, always connected, caught up in one another?' For these authors:

> there are knots of arborescence (rootishness) in rhizomes, and rhizomic offshoots in roots. Moreover there are despotic formations of immanence and channelization specific to rhizomes, just as there are anarchic deformations in transcendent systems of trees, aerial roots and subterranean stems. The important point is that the root-tree and the canal rhizome are not two opposed models: the first operates as a transcendent model and tracing, even if it engenders its own escapes; the second operates as an immanent process that overturns the model and outlines a map, even if it constitutes its own hierarchies, even if it gives rise to a despotic channel.
>
> (Deleuze and Guattari 1988: 20)

This seems particularly pertinent in relation to the ways that events are often partly constituted of enunciations (e.g. narratives, theories, motifs, discourses, slogans, abstractions) that are designed to characterize definitively those events – to territorialize them in particular ways. However, the relation of such enunciations to events are highly complex. Despite themselves, they become 'embroiled' in the event and at once close it down and open it up. Michael and Rosengarten (n.d.) discuss this complexity with regard to the discursive abstractions of the 'gold-standard-ness' (broadly meaning scientific excellence) of randomized controlled trials of pharmaceutical prophylactics for people at high risk of HIV infection. In their analysis, an abstraction acts, and is enacted, in a variety of contrasting ways. First, it is an attractor – a sociomaterial 'aspiration' – toward which an event is moving. It is a specific, virtual prospect which the concrete event is seen to be in the process of realizing. Second, an abstraction is a key element in the concrete making of the event – it is a type of account that contributes to what the event is. Third, in the complex specificity of an event, an abstraction is itself emergent – what that abstraction 'is' is eventuated within and through the specific contingencies and exigencies of the event. Fourth, an abstraction is an ironic element in the problematization of the event – it is a spur to the de-territorialization of the event that lures something 'other', a sort of anti-attractor. In sum, an abstraction at once (i) characterizes an event, (ii) is a component of an event, (iii) emerges through that event, and (iv) precipitates other abstractions that differ from or counter it.

As we shall suggest below, this fourfold schema applies to the particular eventuations of the 3D printing of plastic objects. What amounts to 3D printing is at once commonsensical and fantastical, easy and difficult; and plastic is a matter of opening up and closing down (for a similar discussion of the designed 'thing', see Storni 2012).

The rapid rise of rapid prototyping

Rapid prototyping is a generic name given to a form of additive manufacturing where layers of a material are deposited and fixed on top of one another in order to reproduce a shape that has been determined using computer-aided design (CAD) systems. The materials can vary (e.g. metals such as titanium, or paper, or resin), as can the specific processes of addition and adhesion (e.g. electron beam melting, stereolithography), but here the focus will be on those versions that use plastics (e.g. polylactic acid or polylactide (PLA) and acrylonitrile butadiene styrene (ABS)), and methods such as fused deposition modelling in which a nozzle directs melted plastic (down to the scale of fractions of a millimetre) onto a support platform, where it builds up into the required shape layer by layer.

This technology has been available for some 20 years or so and has been mostly used in industrial and design settings where prototypes (or the components of an artefact in development) can be rapidly produced, and materially examined for fit, aesthetics, usability and various other properties. The key advantage afforded by rapid prototyping is that designers and engineers can quickly and cheaply mock up a given component or artefact and discuss it with the various members of the design or production team. As it was framed in *The Economist* (2011), 'It enables the production of a single item quickly and cheaply – and then another one after the design has been refined'.[3] The upshot is that the process of product design is considerably and cost-effectively accelerated.

Increasingly, however, it seems that this technology is moving toward manufacturing as well as prototyping. Such are the improvements in 3D printing that it:

> is starting to be used to produce the finished items themselves … It is already competitive with plastic injection-moulding for runs of around 1,000 items, and this figure will rise as the technology matures. And because each item is created individually, rather than from a single mould, each can be made slightly differently at almost no extra cost. Mass production could, in short, give way to mass customisation for all kinds of products, from shoes to spectacles to kitchenware.
>
> (*The Economist*, 2011)[4]

Here we have a re-vivified realization of post-Fordist production (e.g. Lash and Urry 1987) where customization of manufactured products can become an inexpensive matter of course. Indeed, it has been suggested that, with the diffusion of 3D printers, there is potentially a movement towards what Craig Allison and his colleagues (n.d.) call a 'hybrid of the consumer/producer dichotomy, a prosumer society … wherein the roles of the consumer and producer merge'.[5]

Having noted this, when following up some of the comments on *The Economist*'s website, we can note that this vision of the future is not an unproblematic one. In reaction to a similar article entitled 'The Printed World: Three-dimensional Printing from Digital Designs Will Transform Manufacturing and Allow More People to Start Making Things' (*The Economist*, 2011),[6] a number of sceptical comments were posted. The first two query the claims made for 3D printing and its supposed advantages over traditional methods of manufacture:

> Nothing really has changed in the last 20 years – it is still a niche as output is to [sic] low, cost high or properties are not met. Too much enthusiasm for something that has obvious inherent problems such as low structural strength and slow production times. Who knows where this will be in 10 or 20 years, but for now it is an amusing sidebar merely. The article does not show the full picture, it misses the advances of foundries, fast milling machines, laser machining etc.
>
> 3D Printing is as close to the market as as [sic] the translation of the Genome project's insights into effective medicines to cure cancer. Watch the hype![7]

These comments reflect how the expectations raised by the enthusiasts for, or advocates of, this or that technology precipitate a negative reaction – the accusation that the claims are fanciful, unrealistic hype (e.g. Brown 2003). Implicit here is a wariness toward the particular performativity of these claims – they are as much concerned with enabling a given future (by generating enthusiasm and, indeed, markets) as depicting it (see Michael 2000b). With these claims and counter-claims, plastic is specifically eventuated (on *The Economist* website) as both a medium for the bespoke manufacture of a multiplicity of objects (de-territorialization, high plasticity in Bensaude Vincent's sense of a material that can be turned into more or less anything) and a material hampered by the limitations of a technology which can, as yet, only produce 'sub-standard' artefacts (re-territorialization, low plasticity in the sense of still failing to realize this promise or prospect of multiplicity).

Here are two more contributions to *The Economist* comments page:

> OK – making progress toward replicators. How's it going with transporters?
>
> we can print out chicken to eat!

In these two cases, there are more and less explicit references to the replicators of *Star Trek*. A replicator is staple technology on the *Star Trek* family of TV and film series, which can fabricate (usually) food (and its receptacles) within a few moments of being verbally commanded to do so. These website comments are, obviously enough, meant to be humorous: they gently and ironically mock the aspirations of those who promote the 3D printer. Yet the term 'replicator' seems to have considerable currency in discussions of the 3D printer. For instance, on a 'dornob: design ideas daily' webpage there is the following headline: '3D Printer + DIY Home Factory = Real-Life Replicator.'[8] On the iTWire (connecting technology professionals) website can be found the recent headline: 'Star Trek Replicator: 21st Century Version Might be 3D Printer'.[9] Finally, the start-up company MakerBot has named its new two-colour 3D printer the 'Replicator'.[10]

Inevitably, the web coverage of MakerBot's innovation has not been shy of referencing *Star Trek* (e.g. 'MakerBot Replicator Beams In'; 'MakerBot Replicator: Out of Star Trek into Your Own Garage'). This connection with the future is certainly present in the MakerBot Replicator's own maker's accounts. For instance, MakerBot's chief executive Bre Pettis has claimed that 'It's a machine that makes you anything you need', and is hopeful that 'if an apocalypse happens people will be ready with MakerBots, building the things they can't buy in stores. So we're not just selling a product, we are changing the future', which includes putting 'MakerBots on the moon (to build) the moon base for us'.[11]

These references to science fiction(-become-fact) eventuate the 3D printer and plastic in a number of ways. The replicator is an abstraction that opens up the eventuation of the 3D printer – that is, the replicator points it toward a particular virtuality (or serves as an attractor or sociomaterial 'aspiration' for the 3D printer), wherein the local production of anything becomes feasible. There is, in other words, the prospect of 'everything-ness' that attaches to the domestic 3D printing of plastic objects. At the same time, the replicator serves to characterize the 3D printer in the here and now, while simultaneously being instantiated by – emerging in its specificity through – the 3D printer. Finally, as we have seen, the 'ingression' of the replicator into the eventuation of the 3D printer also triggers a negative reaction – the associations with science fiction serve simply to underscore the fictional, even fantastical, status of the prospective futures of this technology.

Relatedly, part of the attraction of 3D printers lies in their apparent ease of use – one's CAD designs are seamlessly relayed to the printer through the computer. Indeed, in the various *Economist* comments, it was notable how, despite the criticisms, this ease of operation was assumed – but then many of the posts were from engineers presumably familiar with CAD-like systems. The *Star Trek* idealization of the 3D printer simply builds on this: instructions can be directly conveyed through speech. Oddly enough, it is in *Star Trek* itself where this simplicity of operation is ironized. In an exchange between Tom Paris and the replicator (in the episode 'Caretaker' of *Star Trek Voyager*), it becomes clear that even apparently mundane instruction giving requires an element of skill – instructions need to be highly explicit if they are to be actionable by the replicator. When Tom Paris issues an order to the replicator for tomato soup, the replicator keeps returning with a series of options for different and more specific sorts of tomato soup. The increasingly frustrated Paris ends up growling at the replicator for what seems to him to be the obvious sort of tomato soup – plain and hot. A comment on the same YouTube page parodied the exchange: 'i changed my mind i want pizza now; There are 140 different kinds of …' Tom Paris' annoyance evokes the prospect that, every time one uses a replicator, it would be like entering into a Garfinkelian breaching experiment (Garfinkel 1967), in which what is normally implicit in smooth social intercourse is made frustratingly and disruptively explicit (thus revealing what is implicit). Put another way, specificity in instructions is thus a technically demanding skill.

Home is where the 3D printer is …

The 3D printer, it would seem, is about to become a household item. To read through the various press releases, publicity materials and exhibition reports around these new products is to be left with the impression that their arrival in the home is imminent. This imminence rests on a number of factors: the decreasing costs associated with 3D printers; the convenience of making objects and components that are otherwise unobtainable or hard to obtain; and the ease with which they can be used.

We have seen already how it is claimed, albeit hyperbolically, that the MakerBot Replicator can make anything, but, additionally, it is also asserted that this can be done easily: as it says on MakerBot's website: 'When you get your MakerBot Replicator™, you'll have your machine up and running in no time.'[12] Similar declarations are made for the UP! 3D Printer, which is 'The world's first new standard in personal Desktop 3D printing in price, performance and ease of use!'[13] This sense of the 'ease of use' is reinforced by Dejan Mitrovic's 'Kideville' activity at the Victoria & Albert Museum's 'Power of Making' exhibition, where children were invited to help produce a model city by designing their own house, which was then 3D printed. In sum, in addition to the prospect of 'everything-ness', there is also the prospect of 'easy-ness': anything can be made by anyone, anytime, anywhere.

However, there is much here that is not so 'easy'. First, there is the matter of preparing the designs for 3D printing. As noted, this requires some expertise with CAD systems. In some cases, designs may be downloaded from open-source libraries made available by the 3D printer manufacturer. However, given that one of the key selling points of the 3D printer is customization, one would expect that a facility with CAD is necessary. This seems to be glossed over in many accounts of 3D printers.

Dejan Mitrovic describes the process of Kideville thus:

> I encourage (the children) to split the paper (on which they sketch their picture of a house) into 4 parts so that they think about the different views of the house, so the front view, the side view, the top and then a 3D view. Once they've done that, it helps them understand what their idea is,

> and what they want their idea to look like, and then they move onto the computer where they use special 3D software and quickly design it, mock it up in the CAD and then make a 3D file. After that, I take the file, put it onto the printer and it prints their house out of plastic.[14]

The 3D printer and plastic together become, that is to say, are enacted as, in Bruno Latour's (2005) terms, an intermediary that unproblematically transforms ideas into material objects without deviation or corruption. Yet the 'special 3D software' serves more as what Latour calls a mediator: it realizes individuals' plans in the light of its own particular capacities (as well as in reaction to the capabilities of its users). As such, what seems to materialize is a 'mock up' – an 'estimated' version of the planned object. In this particular eventuation of plastic through the 3D printer, it becomes a material of ready manipulation (a matter of child's play). Yet this is only possible because of the co-presence of a set of skilled practices: drafting and modelling skills (making Plasticine models was also part of Kideville); design skills (thinking about different views of the home); CAD skills; and crafting skills (detaching the plastic objects from the 3D printer, cleaning the object up – see below). At each of these practical junctures, there is a translation of the 'idea' of the house into the emerging plastic house.[15]

In a showcase video for the UP! 3D printer, a child takes her broken robot toy to (presumably) her grandfather. The foot has come off. Grandfather astutely notes that the other foot is identical, so detaches it from the robot and using a digital LCD Vernier calliper takes a number of measurements. The next shots are of the virtual foot taking shape in the CAD system (we are not told how the measurements are transferred from calliper to computer). The final version of the virtual foot is placed within a virtual printer space, and the print command clicked. Two children are then shown looking eagerly at the UP! 3D printer in operation. In a subsequent shot, the completed white plastic foot is shown held in one hand while a second hand using pliers prises a thin sheet of unwanted support plastic from the sole. (A similar but more extensive cleaning-up proccss is shown in another demonstration video for the UP!, where the excess plastic is removed from a ball bearing using both pliers and an awl tool). The foot is then painted black, attached to the robot, and tested for vertical movement by being swivelled up and down on the ankle joint. When the feet are seen together frontally, it is obvious which is the homemade foot (the replacement foot appears lopsided, and it is more matt in colour). The repaired robot is restored to its delighted owner, who exercises it excitedly.[16]

Again we glimpse a complex of complementary skills and capacities necessary for the making of the 3D printed plastic objects. In addition to design, CAD and crafting skills, there are, in this instance, measuring and finishing skills to be taken into account (the latter being especially necessary where the comparatively low resolution of the 3D printer produces an extruded plastic object with a rough surface that requires filing and sanding). In consulting a designer colleague with very considerable experience of 3D printing, it became apparent that a good deal of unexplicated practical expertise and skill was needed for the imagined plastic object of desire to be translated into a printed plastic product.

In summary, we can now draw out a fourfold analysis of (the abstraction that is) 'easy-ness' and its place in the eventuation of 3D printing and its plastics. As such, we can note that 'easy-ness': (i) operates as a tendency that is an attractor toward which 3D printing and its plastics are moving; (ii) serves to characterize the actual operation of 3D printers (this is the way that 3D printers really function); (iii) is also emergent in relation to the contingencies of 3D printing (we see that what counts as a manifestation of easy-ness is up for grabs depending on how an event of 3D printing turns out); and (iv) precipitates a counter-reaction that questions the sociomaterial meaning of the abstract idea of 'easy-ness'.

Concluding remarks

This chapter has attempted to trace the eventuation of 3D printing and plastic in terms of process and plasticity. The argument builds on Bensaude Vincent's (2007) version of plasticity as the pluralization of objects on the basis of the particular capacities (informedness) of plastic – what has here been called 'everything-ness'. In addition to pluralization, there has also been a supplementary emphasis on 'easy-ness' – the making of such a plurality of plastic goods by anyone, anywhere, anytime that has been mediated by the supposed establishment of 3D printing. However, these dual axes of plasticity are, it has been argued, embroiled in the complex processes entailed in the specific and concrete eventuation of 3D printing and its plastic. Developing a fourfold analytic of the event, or rather of eventuation, it has been suggested that these axes serve simultaneously as tendencies or attractors (prospects or virtualities), as actual contributions to the characterization and instantiation of the 3D printing and its plastic artefacts, as characterizations that are themselves contingently emergent through these instantiations, and as prompts for, or evocations of, the 'other' where the axes are problematized (a few things in a few places, rather than everything everywhere).

In terms of the relationship between the 3D printer and sociomaterial respatialization, we propose that the proclaimed collapse of the distinction between domestic and manufacturing or design spaces (and between producer and consumer bodies) might be overstated. Or rather, there is a reconfiguration of these: a space like the 'home post-Fordist factory' rests on the availability of an array of skills – skills that are partly enabled and mediated by the porosity of the home – not least to various knowledges (e.g. access to open-source CAD libraries, websites such as Shapeways[17] on which 3D designs are traded). In other words, the domestic operation of the 3D printer still relies on a 'centre of design expertise' even if that centre is virtual. However, we might also tentatively predict that there will emerge something like a punctuated, circulating expertise or skilfulness as accumulated experience and novel domestically derived designs come to disseminate across the web (and possibly go viral). In other words, we might imagine less a dramatic collapse of the spaces of the 'domestic' and the 'industrial' than a complex shifting interdigitation, one that will be rendered still more complex as the 3D printer itself comes to be 3D printable at home.

Needless to say, the current discussion has been a rather limited one, structured by concerns with the eventuation of plastic in manufacturing, design and domestic spaces (everything-ness), and the relation of such making to skill or craft (easy-ness). Another route might have taken in issues of intellectual property rights (IPR). Might the copying and making of the components of everyday technologies (the cooker knob crops up several times, and is a trial product in one of the UP! promotional videos) infringe the Registered Designs Act, Unregistered Design Right, copyright on 3D Printer Design Files, or the Patents on the technology? As it turns out, in the UK at least, all this is unlikely to be the case (Bradshaw *et al.* 2010). However, if 3D printing were to become highly popular, there is no guarantee that design and manufacturing professional associations will not start lobbying for additional IPR protections. If new IPR measures are put in place, this might conceivably prompt another set of skills (as we see with music file sharing). This possible eventuation of 3D printing and its plastic through the intra-actions of IPR and their infringements are perhaps prefigured in the recent case of the 3D printing of the standard key for Dutch police handcuffs (derived from a high-resolution photograph of the key as it hung from a police officer's belt). The file for the key was put online, rendering Dutch handcuffs potentially useless.[18]

One of the issues that was briefly mentioned above but is otherwise absent from the discussion is the relation of 3D printing to the environmental problems posed by plastic. Given the still emergent character of 3D printing, we can only hint at its implications for the environmental impact of plastic. To start, we can note that *professional* 3D printing can be understood to eventuate plastic in relation to environmental matters of concern in contrasting ways. On the one hand, it can save on the financially and environmentally costly processes of making prototypes through retooling machines on the factory floor; on the other hand, it can encourage designers to 3D print a series of prototypes as the design is progressively – in cheap but environmentally wasteful piecemeal steps – refined (although one of the plastics that is typically used, PLA, is derived from corn starch or sugar cane, and is biodegradable). In relation to the *domestic* 3D printer, there are many additional environmental issues to unpack, such as: learning to do 3D printing entails waste (over and above the almost inevitable mistakes and false starts, freshly 3D printed objects arrive with excess plastic that needs to be removed); there is a 'hobbyist' temptation to make per se, where the act of production is also the act of consumption, and the means of 3D printing (rather than the finished articles) become the end; homemade plastic objects increasingly flow into local (gift) economies as they carry the additional value of 'handcrafted' and become acceptable as presents, keepsakes, mementos and so on. Conversely, as noted briefly above, 3D printing can resource the repairability of plastic objects: no longer the ease of transition from utility to waste.

This potential proliferation of plastic stuff is, of course, grounded in the plasticity of plastic mediated by domestic 3D printing. However, perhaps what is at stake here lies less with 3D printing per se, and more with its (projected) domesticity. The abstractions of everything-ness and easy-ness that attach (albeit, as we have seen, problematically) to the domestic 3D printer serve to individualize production as if it were in principle a good thing when 'production tools become democratized'.[19] Here, to 'become democratized' means to 'become domesticated'. Put another way, this can be understood as another neo-liberal twist in which people are faced with the environmentally charged 'choice' of using 3D printers simply to consume more stuff (make new objects, making as consumption) as opposed to making in order to consume less stuff (by extending the lives of their existing plastic goods).

Notes

1 The author would like to acknowledge the advice and help of Andy Boucher and David Cameron.
2 www.velcro.co.uk/index.php?id=124 (accessed 27 February 2012).
3 www.economist.com/node/18114327?story_id=18114327 (accessed 22 February 2012).
4 www.economist.com/node/18114327?story_id=18114327 (accessed 22 February 2012).
5 Allison, C., Davies, H., Gomer, R. and Nurmikko, T. (n.d.) 'What are the Implications of Personal 3D Printers Becoming Domestically Available?' eprints.webscci.net/8/1/comp_6048_3d_printing.html (accessed 25 February 2012).
6 www.economist.com/node/18114221 (accessed 22 February 2012).
7 Both quotes at www.economist.com/node/18114221/comments#comments (accessed 22 February 2012).
8 dornob.com/3d-printer-diy-home-factory-real-life-replicator/ (accessed 22 February 2012).
9 www.itwire.com/science-news/energy/48686-star-trek-replicator-21st-century-version-might-be-3d-printer (accessed 22 February 2012).
10 www.makerbot.com/blog/2012/01/09/introducing-the-makerbot-replicator/ (accessed 22 February 2012).
11 www.bbc.co.uk/news/technology-16503443 (accessed 24 February 2012).
12 store.makerbot.com/replicator-404.html (accessed 24 February 2012).
13 3dprintingsystems.com (accessed 24 February 2012).
14 www.vam.ac.uk/content/articles/p/powerofmaking/ (accessed 24 February 2012).
15 However, we might well expect that the 'idea' of the object will also come to be shaped by the capabilities of the 3D printer (e.g. Waldby 2000).
16 Both videos are at www.coolcomponents.co.uk/catalog/plus-personal-portable-printer-p-644.html?gclid=CJbRi5WUpa4CFQ8gfAodvmJUPg (accessed 25 February 2012).
17 www.shapeways.com (accessed 28 February 2012).
18 www.shapeways.com/blog/archives/296-German-hacker-3D-prints-Dutch-police-handcuff-key.html (accessed 27 February 2012).
19 Ironically, the home is reterritorialized in distinction to the factory where 3D printers, as opposed to the plastic objects, are made. (Presumably, this will continue

to be the case until 3D printers can replicate themselves.) In any case, this suggests that to trace further the environmental implications of domestic 3D printing is to find that plastic's plasticity proliferates not only plastic objects but also the patterns of de- and re-territorialization of making and consuming, of home and workplace.

References

Akrich, M. (1992) 'The De-scription of Technical Objects', in W.E. Bijker and J. Law (eds) *Shaping Technology/Building Society*, Cambridge, MA.: MIT Press, 205–24.

Akrich, M. and Latour, B. (1992) 'A Summary of a Convenient Vocabulary for the Semiotics of Human and Nonhuman Assemblies', in W.E. Bijker and J. Law (eds) *Shaping Technology/Building Society*, Cambridge, MA.: MIT Press, 259–69.

Barad, K. (2007) *Meeting the Universe Halfway*, Durham, NC: Duke University Press.

Barry, A. (2005) 'Pharmaceutical Matters: The Invention of Informed Materials', *Theory, Culture and Society* 22(1): 51–69.

Bennett, J. (2010) *Vibrant Matter*, Durham, NC: Duke University Press.

Bensaude Vincent, B. (2007) 'Reconfiguring Nature Through Syntheses: From Plastics to Biomimetics', in B. Bensaude Vincent and W.R. Newman (eds) *The Natural and the Artificial. An Ever-Evolving Polarity*, Cambridge, MA.: MIT Press, 293–312.

Bensaude Vincent, B. and Stengers, I. (1996) *A History of Chemistry*, Cambridge, MA: Harvard University Press.

Bowker, G.C. and Star, S.L. (1999) *Sorting Things Out: Classification and its Consequences*, Cambridge, MA.: The MIT Press.

Bradshaw, S., Bowyer, A. and Haufe, P. (2010) 'The Intellectual Property Implications of Low-Cost 3D Printing', *SCRIPTed – A Journal of Law, Technology and Society* 7(1), www.law.ed.ac.uk/ahrc/script-ed/vol7–1/bradshaw.asp (accessed 25 February 2012).

Braun, B. and Whatmore, S.J. (eds) (2010) *Political Matter: Technoscience, Democracy and Public Life*, Minneapolis: University of Minnesota Press.

Brown, N. (2003) 'Hope Against Hype: Accountability in Biopasts, Presents and Futures', *Science Studies* 16(2): 3–21.

Cowan, R.S. (1987) 'The Consumption Junction: A Proposal for Research Strategies in the Sociology of Technology', in W.E. Bijker, T.P. Hughes and T. Pinch (eds) *Social Construction of Technological Systems*, Cambridge, MA.: MIT Press, 253–72.

DeLanda, M. (2002) *Intensive Science and Virtual Philosophy*, London: Continuum.

Deleuze, G. and Guattari, F. (1988) *A Thousand Plateaus: Capitalism and Schizophrenia*, London: Athlone Press.

Fraser, M. (2010) 'Facts, Ethics and Event', in C. Bruun Jensen and K. Rödje (eds) *Deleuzian Intersections in Science, Technology and Anthropology*, New York: Berghahn Press, 57–82.

Garfinkel, H. (1967) *Studies in Ethnomethodology*, Cambridge: Polity Press.

Halewood, M. (2011) *Alfred North Whitehead and Social Theory: The Body, Abstraction, Process*, London: Anthem Press.

Halewood, M. and Michael, M. (2008) 'Being a Sociologist and Becoming a Whiteheadian: Concrescing Methodological Tactics Theory', *Culture and Society* 25(4): 31–56.

Hillman, D. and Gibbs, D. (1998) *Century Makers: One Hundred Clever Things We Take for Granted Which Have Changed Our Lives over the Last One Hundred Years*, London: Weidenfeld and Nicolson.

Lash, S. and Urry, J. (1987) *The End of Organized Capitalism*, Cambridge: Polity Press.

Latour, B. (1992) 'Where are the Missing Masses? A Sociology of a Few Mundane Artifacts', in W.E. Bijker and J. Law (eds) *Shaping Technology/Building Society*, Cambridge, MA.: MIT Press, 225–58.

——(2005) *Reassembling the Social*, Oxford: Oxford University Press.

Lefebvre, H. (1974) *The Production of Space*, Oxford: Blackwell.

Lury, C. (1996) *Consumer Culture*, Cambridge: Polity.

Massumi, B. (2002) *Parables of the Virtual*, Durham, NC: Duke University Press.

Michael, M. (2000a) *Reconnecting Culture, Technology and Nature: From Society to Heterogeneity*, London: Routledge.

——(2000b) 'Futures of the Present: From Performativity to Prehension', in N. Brown, B. Rappert and A. Webster (eds) *Contested Futures*, Aldershot: Ashgate, 21–39.

——(2006) *Technoscience and Everyday Life*, Maidenhead, Berks.: Open University Press/McGraw-Hill.

Michael, M. and Rosengarten, M. (n.d.) *Innovation and Biomedicine: Ethics, Evidence and Expectation in HIV*, Basingstoke: Palgrave, forthcoming.

Mol, A. (2002) *The Body Multiple. Ontology in Medical Practice*, Durham, NC: Duke University Press.

Sennett, R. (2008) *The Craftsman*, New Haven: Yale University Press.

Shove, E., Watson, M., Hand, M. and Ingram, J. (2007) *The Design of Everyday Life*. Oxford: Berg.

Star, S.L. (1989) *Regions of Mind: Brain Research and the Quest for Scientific Certainty*, Palo Alto, CA: Stanford University Press.

Storni, C. (2012) 'Unpacking Design Practices: The Notion of Thing in the Making of Artifacts', *Science, Technology, & Human Values* 37(1): 88–123.

Thwaites, T. (2011) *The Toaster Project*, New York: Princeton Architectural Press.

Waldby, C. (2000) *The Visible Human Project: Informatic Bodies and Posthuman Medicine*, London: Routledge.

Whitehead, A.N. (1929) *Process and Reality. An Essay In Cosmology*, New York: The Free Press.

Part II

Plastic economies

3 Made to be wasted

PET and topologies of disposability

Gay Hawkins

What does plastic do in terms of generating economic value? What is the efficacy of plastic in processes of economic accumulation? How can we think about plastic as an economic agent? These are some of the questions driving this part of the book, and there are many different ways in which they could be answered. One strategy would be to delve into the growing body of research into the evolution of the plastics and petrochemical industries in the twentieth century, and to examine their interrelationships with governments, markets, consumers and the environment. This approach combines political economy with cultural and technological history, and it has produced some excellent accounts of how the *promise* of plastic was realized as both a new industry and a material that profoundly reshaped everyday experience. The focus in this work is on how a steady flow of plastic as raw material exerted a centripetal force on a wide range of industrial processes, organizations and cultural practices – how plastics became central to the development of mass consumption and synthetic modernity (Meikle 1995; Fenichell 1996).

However, as useful as these accounts are for investigating the rise of plastic industries, they can sometimes render plastic a passive object of economic forces. Plastic is represented as something that seems to have an unfolding logic already within it – it is an instrument for capital accumulation. The assumption is that plastic has intrinsic economic values that are realized in processes of industrial research or market application. While there is no question that plastic has been central to massive and diverse processes of industrial and market development, my claim is that its economic capacities are not so much intrinsic as *enacted*. Plastic's economic values have to be elaborated and produced. This is not to say that they are simply socially constructed. Rather, the economic capacities of plastic emerge in specific arrangements and processes, in which the material interacts with any number of other devices – human and non-human – to become valuable.

In this chapter, I investigate how the economic capacities of one particular plastic – polyethylene terephthalate (PET) – have been enacted. I want to analyse how this plastic has acquired the 'character of calculability'. This is a term Callon and Muniesa (2005) use to describe the technical, material and social processes whereby the qualities and value of goods are determined. For

Callon and Muniesa, markets are key sites for the dynamics of calculation – indeed, this is their primary function. However, they are by no means the only site: the qualities or calculability of goods are continually subject to change as they move through various assemblages from design to production to consumption and more. All of these represent specific centres and forms of calculation that elaborate the value of goods or *re*qualify them in various ways. The properties of goods, then, are never fixed; instead, they are continually being enacted in multiple networks and interactions.

PET is the generic name for the bottle form of polyethylene terephthalate, a plastic that forms the basis of the synthetic fibre polyester. Over the last 30 years, production of PET bottles has grown dramatically, and it increasingly has been used to replace glass, metal and other plastic materials (Brooks and Giles 2002). While the PET bottle is often hailed as the container that revolutionized global bottle and beverages markets (Brooks and Giles 2002: 26), in recent years it has also been recognized as a major contributor to the growing environmental burden of plastics. Numerous studies have pointed to the phenomenal growth in plastics waste over the last 30 years. The majority of this plastic comes from food packaging and single-use items such as PET bottles (see Thompson in this volume; Gandy 1994; Melosi 1981). This avalanche of ever-accumulating discarded plastic has put enormous pressure on existing urban waste infrastructures and the natural environments where it often ends up.

Standard economic and corporate analyses of this disturbing waste reality often dismiss it as an unfortunate externality – an inevitable outcome of changed materials or irresponsible consumer practices. However, this dismissal denies the ways in which the calculability of PET has been predicated on how easy it is to waste – on its *disposability*. This has been one of the key defining qualities of this type of plastic. The issues are how this quality has been elaborated and enacted, and the ways in which the material specificity of PET might participate in or resist these elaborations. If, as Callon and Muniesa (2005: 1234) argue, the properties of a good are neither intrinsic nor extrinsic, but are 'co-elaborated', then what kinds of forces and processes have been involved in making PET disposable? How is this disposability enacted as a positive calculation or source of economic gain in some markets and as a negative calculation in others?

In posing these questions, my aim is to situate waste as *immanent* to economic actions rather than as something that follows after them, or that exists as an externality. The tendency in so many accounts of plastics waste is to frame it as a problem that emerges after use – to see the waste realities of disposability as displaced to other times and other spaces.[1] The single-serve PET bottle disrupts this linearity. Like many plastic objects, it appears as rubbish from the beginning. It may have momentary functionality as packaging or as a container, but this is generally subsumed by its more substantial material presence as a transitional object – as something that is *made to be wasted*. As Meikle (1995: 186) says in his account of the rise of disposable plastic objects, 'most remained stylistically rubbish, too anonymous and ephemeral to do more than add to a vague consciousness of an ever greater flow of plastic'. In Meikle's suggestive analysis, 'disposability' emerges as a complex sociomaterial quality central to the movement and apprehension of an increasing number of plastic things since World War II. In relation to PET bottles, it is a set of attributes or qualities that emerge in the processes of making them calculable. However, these processes are immensely variable: they do not simply revolve around the industrial production of bottles or exist in consumption and the cultures of use. They also emerge in the *destruction* of bottles and the industries that have developed to recycle PET – literally to enact disposal. In each of these settings, the 'disposability' of PET means different things and prompts different practices and forms of calculation. In other words, disposability is a shifting quality of the PET bottle, and is continually being *re*qualified in the different arrangements and economies in which it is caught up.

How, then, do we make sense of the multiple economies of disposability? My approach here will be 'topological'.[2] Thinking topologically makes it possible to see how the anticipated future of the single-use PET bottle is folded into the present. This approach understands disposability as an event, and recognizes the extensiveness of this event across multiple spaces from sites of production, usage, waste management and more. More critically, as Michael and Rosengarten (2012) argue, topology does not see time and space as external frameworks but as emergent. As they say, 'transformations of the relations between points are not causal or linear, but open and immanent' (Michael and Rosengarten 2012: 1–2). This makes it possible to consider how the spaces of plastics waste disposal are both distant – in the sense of militantly kept out of market frames – but also proximal, pressing in on processes of production and consumption in ways that can make trouble for the equation of disposability with rapid disappearance. Meikle's reference to disposability as an ever greater *flow* of plastic is also usefully extended via a topological perspective. Rather than see flow as structured by a logic of product life-cycle, topology seeks to map the myriad entities and valuing regimes that become connected in the multidirectional enactments of disposability. In this way, waste – together with the negative value it represents – does not emerge at the end of the product's 'life-cycle'. Instead, it has to be understood as topologically interconnected with the enactment of disposability as the ultimate expression of the convenience of plastic. In other words, the afterlife of the bottle is anticipated before exchange, connecting the value of convenience to the ease with which the bottle is discarded.

In what follows, I investigate three different settings where the disposability of PET is enacted. First, I examine how PET was 'invented' and how the context of industrial research and development (R&D) made this material calculable. Key issues here are how the emergence of the PET bottle can be considered an event, and how the materiality of this plastic became 'economically informed'. Next I explore how PET became a market device – something with the capacity to articulate new economic actions in the beverages industry and to reconfigure drinking. How did this mass plastic become involved in making mass markets and changed consumption practices? Central here is a notion of 'emergent causation' (Bennett 2010), meaning the ways in which the PET bottle acquired the capacity to generate effects – effects that were never purely economic, but also cultural, environmental and more. At the heart of these effects was a distinct enactment of disposability generated in cultures of use. In what ways did consumers interact with the bottle and *re*qualify it as ephemeral, litter, always already waste? In what ways did the bottle suggest such qualifications and encourage drinkers to do new things?

Finally, I examine an example of the literal enactment of disposability – the recycling of PET bottles in Hanoi's 'plastic villages'. Once sites of craft production for the growing city, in the twenty-first century these villages are now focused on recycling the waste of Vietnam's rapid urban and economic development – much of it plastic. In the work of collecting and destroying PET bottles in order to make the crushed pellets available for remanufacture into new plastic objects, disposability is requalified as brute material recalcitrance. In these informal household economies, the calculability of PET is focused on how to dematerialize the single-use object, how to render the incredible durability of PET amenable to new economic actions. In this setting, the PET bottle presents itself as stubbornly *non*-disposable.

These three settings appear disturbingly linear – as if the PET bottle demands a narrative of production, consumption and disposal. This is not my intention. Rather, it is to develop a topological analysis that foregrounds the ways in which enactment is always relational. In this way, the focus is on how each of these specific enactments is folded into the others, and how the PET bottle is the medium through which they become co-present and connect – how the PET bottle becomes a conduit of topological relationships that are continually reframing the play of disposability as material possibility *and* problem.

Inventing PET

The standard story about the invention of PET goes something like this: by the early 1970s, blow-moulded thermoplastic bottles had successfully replaced glass in most household containers for everything from shampoo to detergents. However, application to the beverages industry proved difficult. The thermoplastics used for bottling detergents and other non-drinkable fluids were considered unsuitable for carbonated drinks and fruit juice because the carbonation or acid tended to attack the plastic. This instability led to a range of effects, from deterioration, to explosion of the bottle under pressure of carbonation, to chemical contamination (Freinkel 2011: 172). Beverages therefore presented a vast field of possibility for the expansion of plastic bottle packaging. This was an industry where glass bottles and, from the early 1960s, aluminium cans dominated. While the development of the single-use aluminium can had had a major effect on the growth of markets – particularly Coca-Cola's and Pepsi's decisions to diversify into them in 1967 – there was still an interest in finding a stable plastic that could capture the beverages market.

Then, in the early 1970s, a new form of polyethylene terephthalate was teased out of the lab at DuPont. While this plastic had been around since the 1940s, it was used almost exclusively as a synthetic fibre with the trade name of polyester. The key shift that occurred in the 1970s involved turning this plastic fibre into bottle form. Engineers at DuPont had long been focused on finding a suitable plastic to replace glass bottles. An *object pathway* was in place, but a material innovation was needed to pursue it. The innovation that was developed was a new method of plastic moulding. While bottle production had always involved blow-moulding, the DuPont lab stretched a small test-tube shape of polyester lengthwise *and* widthwise during the blow-moulding process. This conferred remarkable new properties on the material. By realigning the molecules of polyester – by putting them into different relationships – a new substance and object emerged together that was immensely suitable for beverage production. This bottle was very light and virtually unbreakable, it conferred incredible strength on the plastic and gave it remarkable optical qualities to rival glass. As Freinkel says:

> Here was a plastic bottle that was tough enough to withstand all that pressurized fizz but also safe enough to win approval from the FDA. It was as clear as glass but shatterproof and just a fraction of its weight. Its thin walls kept out oxygen that could spoil food contents while holding in that expensive carbon dioxide. The PET bottle was yet another of those pedestrian plastic products that humbly fulfilled a herculean set of demands.
>
> (Freinkel 2011: 172)

In the trade press, 'PET' rapidly became the generic name for the bottle form of polyester. It was attributed to Daniel Wyeth, a DuPont engineer who was hailed as the inventor of the PET bottle. Setting aside the implicit celebration of Wyeth as an individual agent or 'discoverer' of the PET bottle, what the trade press recognized was the research and development context of industrial chemistry. Barry (2005) describes this context as shaped by an 'operational realism' in which new materials emerge out of instrumental and empirical logics. Finding a suitable plastic for beverage bottle packaging applications was the focus of many labs, and this was driven by the potential to capture this lucrative and growing market. The ideal beverage bottle was a singular case, an object that guided research and posed questions to existing plastics and polyethylene bottles.

However, what is problematic about this trade history narrative is the way that it can too easily move from 'discovery' to application – the way in which the PET bottle seems to jump from the lab to the beverages industry in a relatively smooth process of diffusion. Obviously, in the best tradition of science and technology studies (STS), we need to replace application with translation and recognize how the laboratory, the beverages industry and all the bottles out and about in kitchens and laundries and on supermarket shelves generated what Shove *et al.* (2007) call the 'co-production of possibility'. In their account of plastic, they show that the physical and cultural materialization of things feeds back into industrial research practices, investment and innovation, generating 'promise requirement cycles' and socio-technical *co-evolution* (Shove *et al.* 2007: 97). Their point is that 'specific products are as important in shaping the image of plastic (including interpretations of its properties and definitions of performance) as this image is in shaping ideas about potential areas of use and application' (Shove *et al.* 2007: 101).

This important account of co-evolution goes beyond celebratory rhetorics of discovery and application, but it still doesn't quite capture the ways in which materials might *participate* in processes of invention and translation – how they might inform or have a say in what they become. In this framework, plastic emerges out of forces that are largely external to it. While Shove and colleagues are not arguing that materials and products are simply social constructions, their approach is insufficiently attentive to the activity and agency of materials. It is incapable of grappling with the complex ecology of relationships in which a new plastic such as PET was not simply a product of contingent, path-dependent processes of industrial development and translation, but also an event, a novel entity, in which the molecules of plastic revealed new associations and ways of relating. Of course, the molecules didn't do this spontaneously or alone. The emergence of a new material takes place not simply due to things that are imposed on it, but via creative processes of invention in which the material becomes 'richer and richer in information', as Bensaude Vincent and Stengers (1996) argue in *A History of Chemistry*.

Their concept of 'informed materials' suggests a way to extend Shove *et al.*'s (2007) account of co-evolution – or what Callon and Muniesa (2005) call co-elaboration – to the molecular level.[3] For Bensaude Vincent and Stengers, an informed material is one where the:

> material structure becomes richer and richer in information. Accomplishing this requires detailed comprehension of the microscopic structure of materials, because it is playing with these molecular, atomic or even subatomic structures that one can invent materials adapted to industrial demands.
>
> (Bensaude Vincent and Stengers 1996: 206)

Barry (2005) extends this concept of informed materials very productively. In his argument, chemical R&D does more than mechanically reshape or develop new materials for application in various fields: it invents informed materials in which the molecular is always constituted in complex informational and social environments. This topological approach to invention understands it not as progressive evolution or discovery of fixed inherent potential, but as a process of making materials *more and more informed*. In this process, the environment is not external to the material but enters into its constitution. As Barry (2005: 59) says, 'the perception of an entity (such as a molecule) is part of its informational material environment'.

In seeking to understand the emergence of PET, the concept of informed materials is far more incisive than notions of translation for establishing how economic calculability was established for this new plastic. This is because the molecular rearrangement of polyethylene terephthalate into the PET bottle produced rich new information about plastic that was not simply molecular, *but also economic and social*. While there is no question that R&D processes at DuPont were seeking to make the material qualities of polyethylene terephthalate available for economic calculation in the beverages industry, the molecules of polyethylene terephthalate were already economically informed – they were already inscribed with information about plastic's diverse economic and social capacities.

Much of this information came from other thermoplastics. From the mid-1950s, a wide variety of plastics began to be used in containerization and food packaging. These applications helped to establish bottles as one of the key objects where the material and economic possibilities of plastic were enacted. Thermoplastic's soft flexibility, produced through the development of blow-moulding technology, made it the perfect substitute for glass. Here was an unbreakable container that could be shaped in myriad ways. It could also be coloured and squeezed, not just poured. All these qualities highlighted the remarkable 'bottleability' of plastic. They also highlighted the emerging qualities of plastic as a disposable material. As Meikle (1995: 190) argues, abundant supplies of raw materials for thermoplastics and high-volume cheap production following World War II contributed to a gradual shift in perceptions of plastic from being durable to becoming ephemeral. This shift is perfectly captured in an article from a 1957 issue of *Modern Packaging* on 'Molded Plastic Containers':

> The biggest thing that's ever happened in molded plastics so far as packaging is concerned is the acceptance of the idea *that packages are made to be thrown away*. Plastic molders are no longer thinking in terms of re-use refrigerator jars and trinket boxes made to last a lifetime. Taking a tip from the makers of cartons, cans and bottles, they have come to the realization that volume lies in low-cost, single-use expendability … consumers are learning to throw these containers in the trash as nonchalantly as they would discard a paper cup – and in that psychology lies the future of molded plastic packaging.
>
> (n.a. 1957: 120, emphasis added)

This history of the changing uses and meanings of thermoplastics was a crucial part of PET's informational material environment long before it was invented. It showed not only how the molecules of thermoplastics could be manipulated by blow-moulding, but also how these manipulations were implicated in developments in bottle production *and* changing perceptions of plastic. In relation to PET, what *stretch* blow-moulding of polyethylene terephthalate did was form *new relationships* between the molecules of this plastic that actualized the virtual in the form of a new, chemically stable lightweight bottle. The invention of PET, then, was not the realization of a limited set of possibilities inherent within the material. It was evidence of how plastic became richer and richer in information over time through multiple industrial, economic and social enactments.

Developments in post-World War II thermoplastics established *bottles* as one of the key routes whereby material and economic information about plastics' performance became embodied in molecules. The object and the material informed each other. The emergence of the PET bottle represented new material information about molecules and the multiplication of relationships of association between them. PET appeared to enact the 'bottleability' of plastic better than any other plastic. This invention was patterned by historical routes already in place in industrial plastics production and research at the same time as it actualized new properties for the molecules of polyethylene terephthalate and created a new bottle. In this way, the distinct molecular properties of PET need to be considered as *historical* more than as stable physical entities (Barry 2005: 56). Barry explains this key distinction using A.N. Whitehead: 'Whitehead argued that a molecule should be considered an historical rather than a physical entity. In his view a molecule should not be understood as a table or a rock, but rather as an event: "a molecule is a historic route of actual occasions; and such a route is an event"' (Barry 2005: 56).

DeLanda's account of material variability echoes this argument that molecular relationships are historically patterned, informed *and surprising*. For him, the processes of variability or material expressivity reveal the ways in which materials can *have a say* in what they become:

> We are beginning to recover a certain philosophical respect for the inherent morphogenetic potential of all materials. And we now might be in the position to think about the origin of form and structure, not as something imposed from the outside on inert matter, not as a hierarchical command from above as in an assembly line, but as something that may come from the materials, a form we tease out of those materials as we allow them to have their say in the structures we create.
>
> (DeLanda 2005)

Teasing a new bottle form out of polyester was an historical, industrial and molecular process shaped by the logics of extending the economic capacities of plastic. However, as DeLanda shows, materials are not passive in these processes – they express themselves in various ways. In the case of PET, these expressions were not evidence of inherent fixed qualities but of molecular possibilities shaped by the material informational environment that surrounded thermoplastics at the time. The perception that plastic could be disposable was part of this material informational environment. However, this quality of disposability went hand in hand with other qualities that emerged, such as PET's extraordinary durability and clarity. The issue is how these diverse and seemingly contradictory qualities interacted: durability and disposability seem to be opposed qualities? A material that is remarkably tough and unbreakable enables extended use and endurance over time and space. Durability makes commodities mobile, and it enables long periods of storage and extensive supply chain connections. In contrast, disposability implies spatio-temporal transience – an ephemeral material always in the process of becoming waste.

A topological approach shows how the anticipated future of the PET bottle – its status as a throw-away material – was folded into its enactments in markets and consumption. Durability and disposability didn't follow each other in sequence; instead, these qualities informed each other and became complexly interconnected. Making PET calculable meant qualifying it both as durable *and* as intended for a single use. Nor can we consider durability a molecular quality and disposability purely cultural – the consequence of new consumer habits, or 'psychology' to use *Modern Packaging*'s term. These distinctions are impossible to maintain. To cite Whitehead (1929) again, molecules are an historical route of actual occasions, and one of the emerging occasions for thermoplastics following World War II was the perception that high-volume, cheap production made them an abundant and expendable material. This is how they became implicated in the growth of packaging, and how they enabled new economic actions and consumption practices.

Enacting disposability: PET as a market device

In this section, the focus shifts to PET bottles as market devices and to the ways in which disposability is enacted in market arrangements and consumption practices. According to Muniesa, Millo and Callon (2007), a market device is something that articulates economic action – always in relation to other devices. Market devices, then, are objects that *do* things. Whether in a minimal instrumental fashion or a forceful determinist fashion, they are things that act and make others act. This capacity for action is not something possessed by the device: it is an outcome of the distributed agency that emerges in market assemblages. As Callon (1998) argues, a market is not an expression of an economic structure; it is a coordinating device, an endlessly variable assemblage for the calculation of value, the organization of exchange and the distribution of goods, and everything involved in a market arrangement becomes a calculating agent to differing degrees.

This focus on the PET bottle as a market device and calculating agent shifts attention from reductionist notions of cause and effect to the dynamics of 'emergent causation' (Bennett 2010: 33). This inherently topological concept describes the actual processes whereby agency emerges and circulates. Rather than see the take-up of the PET bottle in beverages markets as simply causing market expansion or increased plastics waste, a topological approach maps the multiple effects and spatio-temporalities of the bottle as an actant with emergent capacities rather than fixed or predetermined impacts.[4] Tracking the actions of the PET bottle in beverages markets, then, means paying close attention to the ways in which this plastic became caught up in new relationships and generated effects that were both predictable and unpredictable. It also means investigating how these effects became implicated in further changes via feedback loops. Emergent causation frames the bottle as an object with contingent agentic capacities that are constituted in particular market relationships and uses, but that also *shape* those relationships with shifting degrees of power and effectivity (Bennett 2010: 33).

The take-up of the PET bottle as a new market device in beverages markets had multiple effects. Here I focus on three effects that foreground the complexities of disposability as potentiality and problem. These are the emergence of PET as a mass material; the dynamics of material substitution; and shifting practices of consumer use that revealed a new urban litter problem – namely, increasing amounts of visible plastics waste. In each of these effects, we can see patterns of emergent causation, or the complex processes whereby the PET bottle acquired the capacity to have relational impacts on other things, from plastic production processes to habits of discarding.

In order to make the PET bottle that emerged from the DuPont lab in the early 1970s into a market device, it had to go through various 'qualification trials' (Callon *et al.* 2002: 198). Callon and colleagues describe qualification trials as a sequence of transformations involving different networks and agents that the product singles out and binds together. The effect of these trials, via processes of adjustment and iteration, is to establish and fix the characteristics of the product. In the case of PET, the function of the trials was to stabilize its material qualities and render them predictable – only then could the bottle become a mass-produced market device. These trials involved continued interactions between the plastics industry and both packaging design and bottling companies. Throughout the late 1970s, aspects of PET that resisted full commercialization were researched and revised. The earliest PET bottles were slightly permeable to atmospheric gases, so small amounts of oxygen could enter the bottle. This was enough to spoil the taste of fruit juices and other beverages, and limit their shelf life. In addition, the bottles could not be hot-filled because higher temperatures caused some breakdown of the plastic (Freinkel 2011: 172). All these issues had to be resolved in order to realize the economic capacities of PET. One of the most critical elements in PET's articulation in markets was standardizing it, controlling its material expressivity. To make PET a market device, it had to become a well-disciplined material. This doesn't mean PET was rendered permanently docile – it was still capable of exceeding market regimes and prompting new patterns of emergent causation; rather, stabilizing and standardizing the properties of the plastic were necessary to making it calculable.

Qualification trials underpinned the development of PET not only as a market device but also as a *mass* material. The emergence of PET as a mass material worked in two registers: quantitatively and topologically. The quantitative sense refers to the rapid development of mass production of PET bottles. Thanks to the 'plastics explosion' that had escalated during the 1950s and 1960s in the United States, there were plenty of production facilities making the raw materials necessary for numerous types of plastic. By the 1970s, the plastic packaging industry was extremely well developed, as was the technology for blow-moulding bottles. PET bottles were easily incorporated into existing industrial processes and infrastructures, making high-volume production relatively easy and cheap. By the mid-1980s, PET bottle production had increased phenomenally and glass containerization for beverages was declining. While US commercial production of PET bottles was negligible in 1977, by 1980 some 2.5 billion bottles were being produced, and by 1985 production numbers were up to 5.5 billion units per year (Plastics Academy n.d.).

However, the rapid proliferation of PET bottles also had topological reverberations. Mass production revealed the extensiveness of the event of PET across multiple registers beyond sales figures. This plastic – like many others – generated the effect of being a mass material by being 'ready-to-hand', to use Heidegger's (1962) term – that is, something that involves minimal attention by being available everywhere. Lury (2009) offers an incisive account of the complexities of the term 'mass' in relation to assembling markets. In her analysis, societies predicated on mass production generate a culture of bewildering abundance, which is not stable; rather, it involves a constant flow of products, services and experiences – an ontology of evanescence and transience. As Lury (2009: 73) says, 'mass products draw in the public at the level of the senses, and lead to the restructuring of the very conditions of experience, subjectivity and the body'. In this way, PET can be understood as a mass material that inscribed abundance, ephemerality and disposability in bodily experience. Or course, bodies in mass cultures were already open or willing to be affected in this way. What the rapid accumulation of PET bottles did was enrol those bodies in assembling new markets and drinking practices with a plastic that was both ready to hand and ready to waste.

The emergence and effects of PET as a mass material were also shaped by its capacity to displace other packaging materials in the beverages industry. This process of material substitution was a significant force in the ongoing actualization of PET as a market device. As a new form of packaging, the PET bottle had to be qualified in relation to other packaging in the beverages industry, the liquids it contained and consumers. A key element in this process was how this new packaging interacted with the incumbent materials in beverages packaging – how its qualities became calculated as superior. The usual story of the invention of PET positions the plastic bottle in an almost revolutionary relation to glass. It is represented as a substitute packaging for the glass bottle, which it rapidly pushed aside to become a major competitor with the can and then the market leader. However, the interactions between glass, PET and aluminium were somewhat more complex than just straightforward market substitution. Initially, the characteristics of PET bottles and glass bottles were defined in relationships of similarity. PET was hailed as having qualities that were equal to glass. It was promoted as the first plastic to match the optical standards of glass and to achieve equivalent translucency and clarity. This made replacement of one material with another easier as consumers' expectations of being able to see what they were drinking were not radically disrupted. Like glass, plastic revealed the contents of the bottle to the buyer, preventing unpleasant surprises (see Cochoy 2012). However, unlike glass, the PET bottle was unbreakable: it had durability without fragility and this encouraged new drinking practices.

In relation to the aluminium can, PET was positioned as offering similar possibilities for portability and mobility, but these bottles also suggested a range of new drinking possibilities. For a start, they could be resealed – unlike the spring-pull can – and this made them useful containers for constant sipping over time rather than on-the-spot consumption. These multiple negotiations with other packaging devices in the beverages industry were complex. They involved the dynamics of market positioning that was structured by relationships of similarity and difference. However, as PET became the preferred and dominant packaging material, it emerged as a market device in its own right. It had acquired the capacity to diminish the value of other devices and displace them.

In these patterns of emergent causation – making PET a mass material and rapidly displacing glass and aluminium cans in single-use packaging – we can see how PET's disposability was qualified as a positive force, as a source of market expansion. However, there were unpredicted effects that emerged as PET was taken up in the beverages industry. Specifically, growing amounts of plastics waste rapidly became a new 'matter of concern' in urban waste management. While glass beverage bottles and cans had already established the practice of throw-away containers in recreational drinking, PET prompted new disposal practices – indeed, it gave new meanings to disposability. Discarding glass and cans often required a certain amount of care. Many, of course, entered waste streams, contributing to the massive growth in consumer waste following World War II as the use of disposable objects increased. However, the fact that these forms of packaging had been around for a long time meant that, in a lot of places, public bins, recycling or container deposit systems had developed as an alternative to thoughtless discarding. Both glass and aluminium also presented good opportunities for recycling, as these materials could be reused in the making of new containers.

PET bottles disrupted these arrangements. When they swamped the market industries for recycling plastic were far less developed. PET could also only be *down*cycled because it was impossible to get the same optical clarity using discarded plastic. Each new bottle required raw, not recycled, materials. This absence of effective waste management systems enhanced connotations of PET as *more* ephemeral and disposable than other packaging. As Freinkel (2011: 174) argues, 'the introduction of light, unrefillable PET bottles helped seal the changeover to what the industry calls "one-ways"'. This was a significant effect, and one that Freinkel regards as central to the abandonment of two-way systems in place for other disposable packaging. 'Non-returnables' gave new and very problematic meanings to convenience and disposability.

Beyond the lack of any obvious pathway for the afterlife of the bottle, the affordances of PET as transparent, almost weightless, barely there – unlike the solidity of glass and cans – seemed to lead to an equation of lightness with litter. The PET bottle seemed to invite cavalier waste practices. It gave new meanings to the temporality of disposability and the idea of fleeting utility. So, while PET was rapidly taken up as a replacement for glass, its material performance generated new discarding practices that became increasingly problematic. Its one-way status suggested that it could be discarded with far less concern and care. Its lightness and unbreakability implied that it could be dropped anywhere. It was.

These are some of the *patterns of emergent causation* that reveal how the PET bottle became an actant in the packaging industry – how it acquired the capacity to generate effects. These effects were shaped by various dynamics, such as the topologies of beverages markets, which enrolled the PET bottle in specific ways. At the same time, the materiality of the bottle interacted with these dynamics and made its own suggestions, such as littering and new practices of drinking; it was a device that 'made' consumers do new things. Emergent causation challenges the idea of a simple linear process of industry application or revolutionary material substitution. It foregrounds economic assemblages as patterned by diverse and distributed forms of agency beyond corporate and human intentionality. Via various processes of discipline, massification and interference with other packaging, the *economic expressivity* of the PET bottle emerged and its capacity as a market device was enabled.

The point is that the PET bottle did not enter into the beverages industry with a fixed identity that caused certain effects. Nor was it the passive victim of inexorable economic imperatives or corporate intentions. Rather, it became a participant in already existing markets and, through its interactions and relationships with other devices, helped reconfigure and extend these markets. The PET bottle forged a range of important alliances with continuous-flow oil refining, bottling plants, multinational beverages companies, consumers and more, and as it did so it gained strength to the point where it became capable of profoundly rematerializing beverages packaging. Giles and Bockner (2002) effectively capture the effect of PET as a market device:

> There is little question that the growth of PET packaging has been the big success story in packaging in the 1990s. It now appears that this story will continue to dominate our attention well into the future. PET has demonstrated an unusually strong connection with consumers. For whatever reasons – portability, lightweight, convenience, safety – PET packaging is helping brand owners sell product.
>
> (Giles and Bockner 2002: 26)

Disposing of the disposable

In this final section, the focus is on processes of recycling PET bottles and the multiple shadow realities of disposability. This shift to the ways in which disposable bottles are actually disposed of inevitably reinforces a notion of product life-cycle: it assumes that we have come to the end of the PET story. A topological approach undermines this linearity in several ways. For a start, the claim throughout this chapter has been that the qualities of PET – particularly its disposability – have to be enacted, and in different assemblages these qualities are reconfigured. In invention and mass production, disposability was enacted in relation to high-volume production processes and increasing applications of plastic as a single-use material. In beverages markets, it was enacted in new packaging, drinking and discarding practices. In recycling, disposability is enacted very differently. While recycling processes for PET bottles may seem spatially and temporally separate from production and consumption, the anticipated future of the bottle as rubbish is implicitly folded into these. Disposability is not a sequence of requalifications; it is a dynamic and extensive event mapped out in multidimensional relationships. These relationships have calculative continuities in the sense that there is always a pressure to optimize disposability, to make it a positive value – a source of economic gain. However, these calculative processes are also internal to the spaces in which they emerge; they are not imposed by external forces or coordinates but are emergent (Lury 2009: 77). In this way, topological thinking makes it possible to see how a space for recycling PET bottles has continuities with the other spaces and temporalities bottles inhabit – that it is part of an open network of relationships of variation in which the possibilities for PET are multiple and multiplying.

In order to develop this topological mode of thinking, one site where PET bottles are recycled will be described – what are referred to as 'plastic craft villages' on the urban fringes of Hanoi in Vietnam.[5] These villages were traditionally the site of various forms of craft production, such as weaving, lacquer work or woodwork, which served the needs of the growing city. With mechanization and significant economic growth over the last 30 years, these craft skills have declined and many residents have shifted to servicing the demands of Hanoi's rapidly industrializing economy. One of the new functions identified has been the development of recycling businesses in different villages to deal with the accumulation of specific discarded materials from plastics to paper. These villages are now crucial spaces for the management of the growing amounts of urban and industrial waste that Hanoi produces (Pearse 2010).

Though the focus here is on one place, Hanoi's plastic craft villages are not being used as an illustrative example or as representative of the much larger phenomena of global plastics waste accumulation. They have to be considered in their own right. This is not to deny that they are topologically connected to other PET plastics waste regimes and sites of hazardous plastics accumulation in Vietnam and elsewhere. Instead, it is to stick to the detail and map how processes of recycling PET requalify disposability and generate a complex network of new associations and calculations for bottles. It is to insist, as Law (2004) argues, that the specific and the concrete are where complexity is located.

The key task of the informal recycling businesses in Hanoi's plastic villages is to materially transform PET bottles and other discarded plastics in ways that make them suitable for remanufacture: to convert the wasted plastic object into a raw material. This involves the enactment of distinct labour practices for dematerialization and the calculation of new values. Of course, value is also being generated in the informal economies and translations that shape collection and salvage of the bottles before they are delivered to the plastic villages, but, in the actual villages, the material transformations of recycling have to displace any semblance of the bottles' original object form. Recycling has to render remaindered stuff 'formless in order to be reformed' (Gabrys 2011: 138). This is not a waste-free process. According to Gabrys, the idea that waste can be recycled without remainder is a myth:

> Remainder acquires a duration and delay, circulates through spaces, and undergoes material deformation and transformation but, it persists nonetheless in one form or another … Recycling, in this sense, is never complete and always generates more waste.
>
> (Gabrys 2011: 138)

The disassembly of PET bottles involves an enormous variety of sequenced actions. It involves working with the brute materiality of the plastic, particularly its remarkable durability, in order to transform the object into a material. In these actions, the very affordances that make the bottle disposable, convenient and mobile now become physically recalcitrant and require significant effort to transform. The choreography of recycling involves removing the labels that were stuck on in a microsecond by machine, sorting the lids into different-coloured plastics, cutting off the top of the bottle where the different PVC plastic of the lid leaves a lip, rinsing, and then mechanically crushing, shredding or melting.

To get to plastic pellets or reusable PET flakes is to wrestle with the presence of the bottle as a very tough object, where the calculation of new value is difficult to realize. What makes recycling such a labour-intensive practice, and therefore often concentrated where labour is cheap, is the demands this plastic makes on the human, the ways in which it refuses to cooperate in processes of dematerialization and requalification. In processes of recycling, the power of the PET bottle to demand certain technical and human actions is significant. So too is its power to endure, to enact the temporality of durability by resisting destruction.

This part of the recycling assemblage is not the point at which final waste is averted. Breaking up PET bottles is a filthy, polluting process that is intricately connected to and affected by the diverse flows of waste materials into the villages, and distribution and export out. Many of the PET pellets and chips go back to the Vietnamese plastics production industry as plastic scrap to be mixed with new PET stock in bottle production or to be made into PET fibre, or cheap plastic objects such as coat hangers or chairs (Pearse 2010). Many also go to China for use in the manufacture of carpets and polar fleece. In the plastic villages of Hanoi, we see how entrepreneurial household businesses make the most of the market opportunities that growing amounts of PET and other disposable plastic wastes are generating. There are many successful businesses in the plastic waste villages, and relatively affluent homes where numerous employees are working. However, this success is very exposed to fluctuations in the wider international demand for recycled plastic scrap, as well as local levels of supply and demand. In these plastic villages, discarded PET is helping to produce new economic subjectivities that are certainly less precarious than those of the waste pickers on the streets of Hanoi, who are typically the lowest and most vulnerable labourers in recycling.

However, the idea that recycling represents the straightforward requalification and commodification of plastic waste has to be challenged. This assumes that plastic recycling markets simply express underlying economic or macrosocial structures. In contrast, analysis of Hanoi's plastic villages shows that they operate as 'coordinating devices' – that is, distinct assemblages for the dematerialization of PET bottles and the calculation of new values for the raw materials that they become. In these calculations, plastic presents particular challenges to the enactment of new values. While it is a willing participant in enabling disposability – cheap to produce, light and tough – when it is freed from its role as convenience packaging, it acquires a range of new capacities that generate very distinct challenges to the calculations of new value. It emerges as remarkably *non*-disposable.

Recycling PET bottles is far from straightforward or linear. As the plastic villages of Hanoi show, it is a network of contingent and precarious relationships, choreography and translations that enact various realities. The management of plastic waste through informal household businesses is one of these, but this reality is shadowed by other realities that have significant 'stealth effects' (Law and Lien 2010: 11) – specifically, significant hazards to humans and the environment. While there is no question that 'new' raw materials are being produced, and that new economic capacities for discarded plastic bottles are being *made present*, recycling is also a waste-creating reality. This is evident in the materials that cannot be transformed and are dumped, in the flows of polluted water and plastic sludge running into local rivers, the toxic fumes being released and inhaled as PET is melted, and the millions of plastic chips that blow away in the process. These polluting and occupationally hazardous effects are also being enacted: they are part of the emergent causation of the bottle becoming 'recyclable' and waste. They persist because of poor environmental and occupational health regulations and enforcement in Hanoi, especially in relation to the informal and household sector that dominates in the plastic villages. This is the extensiveness of the event of recycling PET bottles in Hanoi.

Conclusion

This chapter has mapped the shifting economies of disposability enacted by PET bottles. While there is no question that the invention of this plastic object has transformed the packaging and beverages industries, changed how people drink and discard, and contributed to massive increases in the global accumulation of plastics waste, the aim here has been to think topologically rather than through the logic of inexorable effects. The value of accounting for the emergence of PET topologically is that it renders 'effects' as complex interconnections and processes of emergence rather than as predetermined causality shaped by external coordinates. Effects do not express hidden logics or general laws; rather, they emerge in the relational enactment of the PET bottle's various capacities. As DeLanda (2011) argues, capacities are different from material properties; they are what is actualized when things are brought into relation. Capacities, then, are always relational and they are always an event (DeLanda 2011: 4).

Using DeLanda's framework, we can see how the disposability of the PET bottle is a quality or capacity that emerges in its enrolment in packaging, beverages markets, consumption, recycling and more. These various assemblages are structured by distinct spaces of possibility that shape how disposability is actualized: as high turnover, as expendable, as convenience, as one-way, as stubborn material recalcitrance. These assemblages are also connected in heterogeneous relationships and exchanges. Things that seem distant – such as production and consumption, commodity and waste, global and local – 'turn out to be far more promiscuous and can be shown to be in far closer proximity than one might initially imagine' (Michael and Rosengarten 2012: 12). In this way, the single-use PET bottle can be considered a conduit of topological relationships that mixes up plastics waste with plastics production and consumption. The bottle is a medium by which the multiple enactments of disposability become co-present and connect. That shows how the ever-growing flow of plastic moves in chaotic and multiple directions. This was a bottle made to be wasted, and its anticipated future is inscribed in its multiple presents.

Notes

1 It is a standard strategy in corporate campaigns against the introduction of Extended Consumer Responsibility Schemes or Container Deposit Schemes to argue that the waste effects of disposability are completely unrelated to market activities.
2 See the special issue on 'Topologies of Culture' (2012) in *Theory, Culture and Society* 29(4/5) for an overview of current debates.
3 Thanks to Andrea Westermann for this very suggestive point.
4 My use of emergence here is shaped by Harman's (2009) discussion of Latour in *Prince of Networks*. Harman argues that Latour's position is that things are defined by their relations and their outward relational effects on other things, not by their internal composition: 'Latour veers toward a functional concept of emergence: a thing emerges as a real thing when it has new effects on the outside world, not because of any integral emergent reality in the thing itself' (Harman 2009: 158).
5 This section is based on research carried out with the assistance of Warwick Pearse in Hanoi during 2009. I thank him for his assistance.

References

Barry, A. (2005) 'Pharmaceutical Matters: The Invention of Informed Materials', *Theory, Culture and Society* 22(1): 51–59.

Bennett, J. (2010) *Vibrant Matter*, Durham, NC: Duke University Press.

Bensaude Vincent, B. and Stengers, I. (1996) *A History of Chemistry*, Cambridge, MA: Harvard University Press.

Brooks, D. and Giles, G. (eds) (2002) *PET Packaging Technology*, Sheffield: Sheffield Academic Press.

Callon, M. (ed.) (1998) *The Laws of Markets*, London: Blackwell.

Callon, M., Méadel, C. and Rabehariosa, V. (2002) 'The Economy of Qualities', *Economy and Society* 31(2): 194–217.

Callon, M. and Muniesa, F. (2005) 'Economic Markets as Calculative Collective Devices', *Organizational Studies* 26(8): 1229–50.

Cochoy, F. (2012) 'Curiosity, Packaging, and the Economics of Surprise', Charisma Consumer Market Studies website, www.charisma-network.net/markets/curiosity-packaging (accessed 4 October 2012).

DeLanda, M. (2005) 'Uniformity and Variability: An Essay in the Philosophy of Matter', museum.doorsofperception.com/doors3/transcripts/Delanda.html (accessed 7 March 2012).

——(2011) *Philosophy and Simulation: The Emergence of Synthetic Reason*, London: Continuum.

Fenichell, S. (1996) *Plastic: The Making of the Synthetic Century*, New York: Harper Collins.

Freinkel, S. (2011) *Plastic: A Toxic Love Story*, Melbourne: Text.

Gabrys, J. (2011) *Digital Rubbish: A Natural History of Electronics*, Ann Arbor, MI: University of Michigan Press.

Gandy, M. (1994) *Recycling and the Politics of Urban Waste*, New York: St Martin's Press.

Giles, G. and Bockner, G. (2002) 'Commercial Considerations', in D. Brooks and G. Giles (eds) *PET Packaging Technology*, Sheffield: Sheffield Academic Press.

Harman, G. (2009) *Prince of Networks: Bruno Latour and Metaphysics*, Melbourne: Re.press.

Heidegger, M. (1962) *Being and Time*, trans. J. Macquarrie and E. Robinson, New York: Harper & Row.

Law, J. (2004) 'And if the Global were Small and Noncoherent? Method, Complexity and the Baroque', *Environment and Planning D: Society and Space* 22: 13–26.

Law, J. and Lien, M. (2010) 'Slippery: Field Notes on Empirical Ontology', unpublished essay, heterogeneities.net/projects.htm (accessed 23 June 2012).

Lury, C. (2009) 'Brand as Assemblage', *Journal of Cultural Economy* 2(1): 67–82.

Meikle, J. (1995) *American Plastic: A Cultural History*, New Brunswick, NJ: Rutgers University Press.

Melosi, M. (1981) *Garbage in Cities*, College Station and London: Texas: A&M University Press, 1981.

Michael, M. and Rosengarten, M. (2012) 'HIV, Globalization and Topology: Of Prepositions and Propositions', *Theory, Culture and Society* 29(4/5): 1–23.

Muniesa, F., Millo, Y. and Callon, M. (eds) (2007) 'An Introduction to Market Devices', in F. Muniesa, Y. Millo and M. Callon, *Market Devices*, Oxford: Blackwell, 1–12.

n.a. (1957) 'Molded Plastic Containers', *Modern Packaging* 31(1): 120–23, 240–41.

Pearse, W. (2010) 'A Look at Vietnam's Plastic Craft Villages', ourworld.unu.edu/en/a-look-at-vietnam's-plastic-craft-villages (accessed 23 June 2012).

Plastics Academy (n.d.) Hall of Fame website, www.plasticshalloffame.com (accessed 1 October 2012).

Shove, E., Watson, M., Hand, M. and Ingram, J. (2007) *The Design of Everyday Life*, Oxford: Berg.

Whitehead, A.N. (1929) *Process and Reality: An Essay in Cosmology*, New York: The Free Press.

4 The material politics of vinyl

How the state, industry and citizens created and transformed West Germany's consumer democracy

Andrea Westermann

British historian Richard Evans recently restated what has become common sense among scholars of German history:

> Germany today is the product of the 'economic miracle' of the 1950s and 1960s, when rapid recovery from the devastation of the war fuelled a prosperity and stability in the German economy that finally reconciled ordinary Germans to the virtues of democracy.
>
> (Evans 2011)

Yet the positive relationship between economic prosperity and democratic conviction foregrounded by Evans has too often remained an assumption, corroborated with hindsight by West Germany's peaceful trajectory through the decades of Cold War. The 'acceptance of democracy as a life form' persists as a 'core problem' of research (Jarausch 2005: 71). I argue that focusing on vinyl helps to describe and explain exactly how West Germans appropriated the values of democracy. Polyvinyl chloride (PVC), better known as vinyl, was the first fully synthetic thermoplastic produced by German industry in the 1930s, and the one most widely used in West Germany until the mid-1970s.[1] This chapter traces its development and uses before and after World War II. Before 1945, vinyl came into being on an industrial scale within the context of the National Socialist economy of autarky and war. After 1945, it provided infrastructural support, generated myriad commodities and produced a distinct aesthetic for the reconstruction of post-war West Germany. Together with other thermoplastics, vinyl became the iconic material of the social market economy – or, as the accompanying political culture has aptly been called, of West Germany's consumer democracy (Carter 1997: 24).

In the twentieth century, 'consumption became a political project intimately bound up with the state' (Hilton 2007: 66). Following this trend, the West German state sought to achieve social coherence and gain the loyalty of its citizens by promising mass consumption and individual prosperity. Under the auspices of the Allied powers and the US European Recovery Program in particular, West Germany created its own version of liberal democracy. Based on the social market economy, it offered its citizens a form of social inclusion that was not overtly political but sought to give an economic connotation to the democratic principles of individual participation in society and electoral freedom. Citizens were perceived as consumer citizens who would engage in the sphere of consumption rather than in the political arena (Merritt and Merritt 1970: 43; Grosser *et al.* 1996: 97; Schildt 2000).

As with any other artefact, the focus on vinyl pays off methodologically because it makes visible connections between the micro-level of individual action and the level of collective or organizational action (Krohn 1989; Latour 1983; Shove *et al.* 2007: 93–116; Thévenot 2001). However, this mediating capacity does not account for all of vinyl's power of historical explanation. Vinyl's exemplarity regarding the analysis of a consumer democracy also comes from its basic material properties – in particular, its malleability and synthetic origins. As a thermoplastic, vinyl melts and becomes malleable when heated; thus, it came close to fulfilling the dream of infinite potential uses and unlimited forms that the plastics industry had promoted since the introduction of celluloid in the 1870s (Friedel 1983). Taken together, vinyl's plasticity and its chemical creation captured what high modernity expected from technology at large: a world freed from the material restrictions that nature traditionally imposed on humanity. By implication, we would also have a world freed of scarcity, a world of plenty. From the beginning, modern plastics were produced, used and praised to help meet the needs of a mass culture in the making. In Imperial Germany, 'surrogates', as they were called, lowered the costs of production of desired commodities by means of imitation, making luxury goods affordable to many (Koller 1893: vii). The National Socialist economy of autarky and war was an economy of scarcity in many ways, and, as a rule, *ersatz* products, already known from World War I, figured prominently as indicators of want (Tooze 2006: 145–65). Against this backdrop, I suggest that, even in the Third Reich, vinyl was found on the side of abundance. Grounded in scientific progress and abundant supplies of coal, its production forged a path to the systematic and continuous extension of material resources. National Socialist politics increased and refined the know-how involved in the mass production of vinyl. It also relied on the cultural link between plastics and mass consumption.

However, to some people, vinyl was not a true indication of how science and technology facilitated abundance in the twentieth century. Since the 1950s, it had given them reason to criticize what they believed was wrong with the rise of mass culture. Thermoplastics became a symbol of the downside of affluent society. Malleability meant shoddiness, falsity and veneer. Critics condemned thermoplastics for their superficiality and their cheap copying of expensive materials and craftsmanship, all contributing to the mere 'façade luxury' of the West German 'economic miracle' (Kogon 1964: 104). They felt that Ludwig Erhard's 'freedom of consumer choice' was only a 'surrogate of individual freedom' (Freyer 1957: 112). Also, vinyl's omnipresence and non-biodegradability made waste problems tangible, and nowhere was this more evident than in the growing piles of discarded plastic packaging. Furthermore, vinyl's chemical composition created unintended health hazards. As the critiques accumulated, overlapping and corroborating each other, vinyl became a powerful source for the rise of consumer activism. The widespread disappointment and concern about vinyl had a mobilizing effect (Hirschman 1982). People started to re-evaluate the very foundations of West Germany's consumption-based political culture.[2] In this way, vinyl served both efforts: a consumer democracy conceived by the state in the immediate post-war period and based on depoliticized notions of integration and abundant choice, *and* the emergence of citizen activists who re-politicized consumer democracy from the late 1960s onwards. The mechanisms of these two modes of participation are explored in the latter sections of the chapter.

Finally, do plastics drive history? I take up this methodological question in the concluding section in order to link my findings to the overall issue of the book: the material politics of plastic.

The 'better' *ersatz* materials: thermoplastics in an economy of autarky and war, 1933–45

Thermoplastics were mainly produced at chemical giant I.G. Farben, Germany's largest corporation in the interwar years. In 1927, 13 per cent of its sales volume went into research and development (R&D) and half of that was spent on developing new products such as thermoplastics. Even after cutting back half of its R&D expenditure in 1931, I.G. Farben continued to invest 6 per cent of its sales in research – a level unsurpassed by foreign rivals. Plastics' share of the R&D budget shows that this field 'counted among the most important new research sectors between 1932 and 1935' (Abelshauser 2004: 237, 267–68; Plumpe 1990: 337; Stokes 2000b: 405). In the early 1930s, the I.G. Farben laboratories achieved impressive results. Vinyl turned out to be the most interesting plastic because it was suitable for many purposes (Plumpe 1990: 154, 333). Yet the first items made of vinyl were presented in modest terms at a meeting of the newly established I.G. Farben committee on plastics in October 1930 (Hoechst Company Archives 1966: 42). An engineer had made some vinyl plates. Without considering them a 'big success', he emphasized that compression moulding of PVC, 'which to many of us had seemed a hopeless substance', had worked after all.[3] By the mid-1930s, synthetics were becoming ideal vehicles to comply with National Socialist decrees of self-sufficiency. Two documents show how chemically produced *ersatz* materials began to articulate the decrees: an I.G. Farben exposé on the 'new targets and tasks with respect to I.G. Farben's politics of R&D and production resulting from the developments in Germany and on the world market', drafted in 1935 (Hayes 1987: 128–30; Lindner 2008: 272–77), and Hitler's Four-Year Plan memorandum of 1936 (Treue 1955; for an English summary Tooze 2006: 219–22). I.G. Farben's internal memorandum reminded readers that the changed world market in the aftermath of World War I had left the corporation with only half of its export revenues due to German reparations, new international competitors and increased state investments in these industries as a result of global rearmament efforts. The domestic market, in contrast, now made up 60 per cent of I.G. Farben sales, and promised a growing demand for synthetic substitutes because most of them 'may also become of importance militarily'. The analysis stated the need for a private-sector company to 'adjust to changes in foreign and domestic demand which have already occurred and which are likely to be even more extensive' (Lindner 2008: 264). The company readily turned its back on international markets in exchange for a lucrative domestic economy of autarky and war.[4]

While the document testified to I.G. Farben's self-mobilization, Hitler's memorandum of 1936 called for the general mobilization of industry. In keeping with the future-oriented language of National Socialism, combining promise and command (Ehlich 1989: 23–24), it pictured a pending emergency situation to which the launch of a Four-Year Plan was a response. 'Catastrophe was near', Hitler proclaimed. The 'final destruction' of Germany loomed large. He famously concluded that 'the German economy must be fit for war within four years' (Treue 1955: 205, 210, 220). Hitler saw no point in merely repeating that Germany lacked raw materials and foodstuffs. Referring to industry's flexibility, he polemically called for its representatives to join in Germany's struggle for survival and provide creative solutions:

> Either we possess today a private industry, in which case its job is to rack its brains about methods of production, or we believe that it is the government's job to determine production methods, and in this case we have no further need of private industry.
>
> (Tooze 2006: 221)

Hitler modelled the solutions as technical ones and explicitly called for the 'rapid', 'determined' and 'ruthless' development and use of synthetic technologies. In radical jargon, he dismissed the issue of economic viability as 'totally irrelevant' (Treue 1955: 208).

Hitler's Four-Year Plan memorandum specifically mentioned synthetic rubber, synthetic fuel, synthetic fat and light metals. Yet thermoplastics also had a role in the material politics of National Socialism. It was often heard that, in the case of mobilization, 'we have to secure also the raw substances and additives, all available in Germany, for plastics like polyvinyl chloride or polystyrol'.[5] This remained true even though thermoplastics' share of I.G. Farben's overall investments in R&D and general plant construction declined after 1940 and never competed with two of the most ambitious *ersatz* programmes of both the state and the I.G. Farben corporation, synthetic (air) fuel and synthetic rubber (Birkenfeld 1964; Lorentz and Erker 2003; Plumpe 1990: 338; Tooze 2006: 449). Exact numbers are hard to come by, but a graph prepared by the Imperial Office of Economic Advancement in 1940 visualized the production of thermoplastics. It showed that vinyl and its various compounds that came under the name 'Igelite' remained the most widely produced items within the emerging group of thermoplastics.[6] 'Igelite' was the plural form of 'Igelit', the I.G. trademark for polyvinyl chloride, but other names for vinyl also existed. In different compositions it was sold as Mipolam, Astralon or Luvitherm. In 1936, a total of 267 tons of vinyl were produced, compared with 1,467 tons in 1938 and 11,000 tons in 1940. The plan was almost to double production by 1942. The war stimulated the actual use of vinyl, as was confirmed by an I.G. Farben manager. He explained that, while the vinyl production plant built in Bitterfeld in 1938 had 'started out producing for the stockpile', 'things changed for the better for vinyl, almost from one day to the next, because armament efforts led to a lack of traditional materials' (Westermann 2007: 317). By 1944, a vinyl quota was officially allocated to 11 commissions set up within the war regime of economic dirigisme.[7] Vinyl substituted a variety of materials: glass, leather, rubber and non-ferrous metals in the cable and pipe fitting industries, as well as in the chemical and textile industries.[8] At the end of the war, considerable experience with polymerizing vinyl on a large scale, and with wide-ranging technical applications of the thermoplastic, had been developed. For instance, innovations regarding polymerizing procedures had been made, which dramatically increased the quantities of vinyl produced at one time (Westermann 2007: 76).

The early uses of vinyl not only yielded technical know-how, but also forged vinyl's symbolic association with growing mass consumption. Plastic experts justified the significance of vinyl by celebrating its properties. To the developers, vinyl and other thermoplastics were not like other *ersatz* materials: they were 'better' *ersatz* materials for various reasons. Recycling of materials and the long-term planning of raw material supplies simply perpetuated traditional practices of economic thrift. Thermoplastic research, in contrast, revealed untapped opportunities. Unlike synthetic fuel or food, these surrogates were not confined to substituting particular natural resources, but lent themselves to a variety of industrial uses, driving innovations in the industries processing them into finished goods. They were suited not only to coping with shortage but also to achieving post-war prosperity. By 1932, I.G. Farben and its partners had already produced prototypes of virtually all the vinyl commodities that would be in use by the late 1950s. 'Soap boxes, containers, balls, bowls … jam jars, other packaging cans, cups' were extruded out of hard and soft PVC films, and the list was getting longer, including 'bed linen, posters, passport covers, dials and other printed foils, windows, shields for gas masks, records, movable type for letterpress printing'.[9] In the early 1930s, those researching and developing vinyl expected a flourishing mass culture to emerge. Even as the Nazi economy of autarky and war started to restructure both market realities and business organization in the mid-1930s, plastics scientists and engineers emphasized the 'promise' rather than the 'command' side of National Socialist propaganda: 'It is up to the German chemists to fine-tune the material properties and make them, according to requirements, solid or liquid, soft or flexible, hard or brittle', declared the editor of the journal *Vierjahresplan* and head of the chemical section of the Four-Year Plan organization, Johannes Eckell (1937: 283). Plastic pioneers diverted notions of factual constraint and *ersatz* towards ideas of innovation (the name 'Neustoffe' was suggested to replace *ersatz*), feasibility and variety, including peacetime uses (Leysieffer 1935: 6).[10] This sense of long-term change and creativity associated with their objects of study was well founded: they felt they were part of a new and exciting field of research: macromolecular chemistry (Furukawa 1998; Westermann 2007: 60–80).

National Socialism started to use vinyl for the production of finished consumer goods: raincoats, shower curtains, shoe soles and shoes, or the inner walls of the planned 'people's refrigerator' (Westermann 2007: 49, 80). At times, politics explicitly strengthened the link between thermoplastic goods and the opportunities for the good life these commodities promised. The official market launch of vinyl took place in 1937 at the Düsseldorf exhibition 'Reichsausstellung Schaffendes Volk'. The event was intended to display the regenerative effects that the Four-Year Plan was having on the German economy (n.a. 1937a). The idea that the state could provide access to goods and services for the many, thus finding popular resonance and emotionally binding its citizens to the larger political and social order, was one the National Socialist regime took seriously. With the 'people's community' (*Volksgemeinschaft*), it offered a racist vision of an affluent mass society. For the purposes of persuasion and consent, National Socialists explored and used the logics of mass consumption – even though historical research is divided about the extent to which the regime was able to provide for consumer capitalist climate and infrastructure (Abelshauser 1998; König 2004; Spoerer 2005; Wiesen 2011). At any rate, the *Reichsausstellung* was an extraordinary publicity campaign for synthetics of every kind: the thermoplastics Igelit and polystyrol, acrylic glass, the half-synthetic staple fibre 'vistra', the synthetic rubber 'buna' and synthetic fuel.[11] Foreign observers were impressed with 'the extent to which the Germans are developing the mass production of synthetic materials, for the perfection of the finished product and for the many uses to which they are putting them' (Tolischius 1937: 6). Yet thermoplastics were modern everywhere. At the World Fair in Paris, held in the same year, prizes were awarded to I.G. Farben's vinyl.[12] To summarize: in National Socialist Germany, thermoplastics' *ersatz* identity complemented the ideological imperative of coping with the adversities ensuing from a permanent state of exception. Yet the use of thermoplastics had been more than a mere emergency measure. The malleability of thermoplastics carried a surplus of meaning fuelled by older visions of surrogates on the one hand and the advances of macromolecular chemistry on the other.

Vinyl as a medium of consumer democratic integration and communication

This surplus was activated after the war. In West Germany, vinyl's plasticity rapidly came to stand for the consumer citizen's freedom to choose. Time and again, texts and talks promoting the regime of a social market economy asserted that 'democracy and free economy went together naturally' (Löffler 2002: 576). In 1956, Alfred Müller-Armack, who coined the concept 'soziale Marktwirtschaft', explained that the programme combined elements of economic and social theory, and would 'through further expansion, raise the standard of living of all social groups' (Müller-Armack 1956: 392). Purchasing power and consumption made almost everybody feel as if they were 'already sharing in the abundance and luxury of the everyday'; most importantly, as sociologist Helmut Schelsky stated in the mid-1950s, 'participation is considered a civil right' (Schelsky 1965: 340). In technical terms, the exchange of commodities served as a vehicle of social inclusion: it tied consumer citizens to a social course of action. The post-war advocates of mass prosperity perceived this form of affiliation as a sorely needed objectification (*Versachlichung*) of political belonging (König 1965a: 485; König 1965b: 468). In their view, the emerging West German consumer democracy differed – as it should – from the stridently politicized ideal of the National Socialist *Volksgemeinschaft*.

The first testimony to vinyl's role in the changed circumstances of post-war Germany came from amidst the rubble of the bombed-out city of Berlin. As early as 1946, architect Hans Scharoun, head of the urban planning department in Berlin, praised the 'light, purely chemically produced material which has until now not been used for building purposes'.[13] This material would be able to ease the shortage of mass housing after the devastation of World War II. In the ruins of the Berlin Royal Palace, he exhibited models of 'mass-produced full-plastic houses' made entirely of vinyl (Tschanter 1946). The Allied authorities had specified the design and the technical criteria for various house types called 'America', 'England', 'France' and 'Russia'. Scharoun himself had at least projected the fifth house type, 'Germany'. The models were displayed on a 1:5 scale. Different types of vinyl and vinyl compounds were used for the walls, the roof, the windows and doorframes, the pipes and the electrical infrastructure (Böttcher 1946; Hausen 1947). The three functions of stone walls – support, cubature and insulation – were distributed on three plastics elements: a frame construction, an outer skin and an insulation layer. The total weight of the 65 m^2 house amounted to 3 tons, which was 40 times lighter than traditional houses.

Given the acute housing shortage Europe faced in the immediate aftermath of the war (Diefendorf 1993), it seemed more than reasonable that the construction industry should exploit the possibilities of rationalization. Yet, although Berlin would have been an obvious site for installing the prefabricated houses, Scharoun transcended the desperate situation of his home.

The thermoplastic houses promised to 'solve the housing problem the whole world is grappling with'. Inadvertently perhaps, the architect presented universally useful engineering work 'made in Germany' as political reparations. He hoped to correct the dominant image of a destructive Germany. The emphasis placed on design and the choice of five types of house pointed beyond the idea of temporary emergency shelters. The post-war meaning of vinyl's plasticity, denoting customization and plenty, started to take shape in the house models. Scharoun eclipsed the imperative of tackling wartime constraints that consumers might associate with *ersatz* materials by foregrounding the idea of a 'peaceful reconstruction of the world'. In this view, vinyl promised to ensure global prosperity, diversity and individual economic recovery.[14] Similarly, at the Nuremberg Trials against the managers of I.G. Farben corporation in December 1947, the defence lawyer of the Ludwigshafen Director Otto Ambros gave the malleability of thermoplastics a touch of universal humanitarianism:

> It becomes evident here that, due to almost limitless possibilities in the choice of primary materials and of methods, it is possible to give the final product any desired quality that will best suit it to human needs.
>
> (Trials of War Criminals 1953: 279)

The wording might have been chosen with care in a trial the indictment of which accused the defendants 'of major responsibility for visiting upon mankind the most searing and catastrophic war in human history' (Trials of War Criminals 1953: 99). The lawyer presented his client as a visionary protagonist of the dawning 'age of plastics' (Trials of War Criminals 1953: 279). The attractive formula, created in the United States (Meikle 1995: 73), soon gained traction in Germany. In 1950, a plastics scientist justified the new designation of the epoch by asserting that synthetics were 'the material that best suited our fast … and highly diversifying time' (Stoeckhert 1950: 279). Thermoplastic malleability promised a free hand in dealing with the challenges of the present era, rapid change and individualism.

Thermoplastics complied with and furthered the strategy of consumer-targeted diversification like no other mass-produced material before them. They were capable of imitating virtually everything and adopting just about every material quality. How, though, was the plastics industry to explain this good news to industrial manufacturers and consumers? How could it explain exactly what enabled thermoplastics' versatility? In 1952, an educational exhibit in Düsseldorf, extending over a surface of 800 m^2, provided some answers. It was part of the plastics trade fair K'52, the first major event organized by the plastics industry after the *Reichsausstellung* in 1937.

Visited by 165,000 people, the K'52 Trade Fair was an elaborate scheme to promote plastics and its experts (n.a. 1952). The 'Alphabet of Plastics' exhibition was dedicated to the physical and chemical properties of thermoplastics, the 'matter without structure or texture as we know it' (Saechtling 1952).[15] It offered a five-step exercise in scaling down from the 'diameter of the earth to the atomic nucleus' in order to attune visitors to the macromolecular world of polymer matter. Visitors learned that the 'manifold appearances' of plastic were thanks not only to the fact that different substances – such as melamine, vinyl or polyethylene – were grouped together, but also that the diversity of synthetics was the result of molecular manipulation. Through combinatory methods, atoms and molecules were linked together in macromolecules that could yield any desired property. The exhibition was made with some care, and did not shy away from the complexities of the subject. Vinyl was chosen as the exemplary object of demonstration. A three-dimensional 'calotte model' revealed its molecular structure and explained the properties of the material. The model consisted of spherical caps tangent to each other. Also known as a 'space-filling model', it was a physicist's contribution to ongoing plastics research rooted in organic chemistry (Westermann 2007: 102–10).

According to its creator, Herbert Stuart from Hanover, older models representing the atoms as whole spheres or balls gave a distorted picture of how the macromolecules extended in space. Due to the overlapping of the spaces taken up by the electrons in each atom of a molecule, the relative size of the different atoms and their 'true spatial arrangement' could be depicted only if the material entities representing the atoms had the shape of spherical caps instead of whole balls or spheres (Stuart 1952: 101–2). Stuart had marketed his calottes as model kits since the 1930s, as did Linus Pauling in the United States (Francoeur 1997). In the second half of the twentieth century, these space-filling models became standard practice in science and science education. At K'52, industry presented thermoplastics in ways that were fascinating to scientists and the public alike. The vinyl model was the first space-filling model of polyvinyl chloride ever produced. The representational accuracy of the exhibition highlighted the message directed at the visitors: that the plastics industry was guided by scientific diligence and reason in whatever it did, whether regarding economic decision-making or offering technical solutions to other branches of industry.

West Germans quickly incorporated plastics in their everyday lives. The per capita consumption of all plastics increased from 1.9 kg in 1950 to 15 kg in 1960. By 1962, West Germans were the leading consumers of synthetics in Europe, ahead even of the Americans at 21.4 kg versus 18.1 kg per capita (Reuben and Burstall 1973: 35). Three industries most conspicuously worked towards mass prosperity and all three were involved in the growth of plastics consumption: the construction industry, the industry of (electronic) mass consumer goods, and the packaging industry. All of them used extruded or moulded vinyl, and hard and soft vinyl foils, to reconstruct the materials and spaces of daily life.

Plastic packaging, in particular, facilitated mass consumption. Vinyl allowed for improved methods of logistics, storage, conservation and sale. The new ways of handling and distributing commodities in retail and wholesale were not only based on plastic containers and plastic bags, but also required an improved stackability of goods, achieved by material innovations like shrink-wrap.[16] Shrink foils were part and parcel of palletizing, and standardized pallets were used for efficient transport and storage; the pallets, in turn, were only one element among others in the revolutionizing of the global value chain (Dommann 2009; Girschik 2010). The plastics industry estimated that 360,000 tons of plastics had gone into the packaging sector in 1969; the estimates for 1972 were already much higher, namely 969,000 tons (Stoeckhert 1975: 241).[17]

The built environment was increasingly being made of thermoplastics: building façades, the quilted walls of cinema theatres, tablecloths, public transport upholstery and much more (Westermann 2008). These interiors were examples of 'small consumption', already affordable for many and envisioning a full-blown prosperity yet to come (Betts 2001: 185–217; Westermann 2007: 190–201). They also represented the normalcy that found its expression in smooth, undamaged surfaces – this was in strong contrast to the rubble and destruction of war.

Product designers and architects took advantage of vinyl's aesthetic flexibility. Art leather made of vinyl was used prominently in urban transportation systems, the national railways and cars. J.H. Benecke, a leatherworking company from Hanover, had been substituting vinyl for leather or fabric since the early 1940s, and became an important manufacturer of vinyl films a decade later. Guided by concerns about what constituted the typical feel and texture of leather, J.H. Benecke's chemical engineers turned to the material sciences and developed methods of material testing in an attempt to appreciate the consumer's perspective and experience. In order to achieve technical standards for vinyl processing and perfect their art of imitation, they aimed to quantify people's ideas about usefulness and their intimate knowledge of traditional materials (Westermann 2007: 172–76). Vinyl also helped realize and revive a sober aesthetic of functionality. In the post-war period, the non-grandiose aesthetic of functionality soon became linked to West Germany's political culture. The Bonn parliament was modelled on these standards in 1949. Vinyl flooring was considered to be the adequate foundation of the newly established parliamentary routine. Designed as a 'space for dialogue', the architecture and interior design of the building sought to resonate with 'the human voice'. The voices should neither sound cold or hard, nor be swallowed by 'the dusty media of representation deadening any tone with stucco, carpet, curtain, cushion'. Materials were chosen from the 'sober and strict media of a technical age' instead: 'iron, aluminium, painted walls, synthetic flooring' (Schwippert 1951: Preface, 68).

However, the extensive use of plastics did provoke highbrow critique. This opposition expressed elitist scepticism about West Germany's political road. Both left-leaning and conservative intellectuals aired their suspicions of the poor power of aesthetic judgement their fellow citizens exhibited by buying bright and colourful plastic things. What did these choices reveal about their political preferences and the political climate? Author Hans Magnus Enzensberger reviewed the mail order catalogue sent to every household by the pioneer in this business, Neckermann, in 1960. The catalogue, he concluded, was more than a product of commercial calculation: 'It is the result of an invisible plebiscite.' Arguing with an instrument hardly provided for in West Germany's Constitutional Law, Enzensberger foregrounded the dangerous superiority of the sovereign. He deconstructed the plebiscite's outcome. The majority of West Germany had voted for 'petty bourgeois hell'. Everywhere, the writer detected 'reactionary garbage, hidden behind polished polyester surfaces', and 'surrogate art'. Unfailingly, the plastic products evoked traditional oppositions: dirty vs. clean, authentic vs. surrogate, surface vs. substance (Enzensberger 1962: 141, 149). Wolfgang Koeppen's protagonist in *The Hothouse* (2002), a novel about the early years of the Federal Republic of Germany, is equally explicit in his refusal to accept the political status quo. A member of the Bonn parliament, he fiercely rejects the contract offered by consumer democracy: 'Don't take part, don't participate, don't sign on the dotted line, don't be a consumer or a subject ... Ascetic *Keetenheuve. Keetenheuve disciple of Zen.*' The shop window family he passes in a night-time Bonn street disgusts him because 'the grinning father, the grinning mother and the grinning child' staring delightedly at their price tags seem to embody the consumer citizens he abhors: 'They lead a clean, cheap, ideal life. Even the provocatively thrusting belly of the fashionable doll, the little slut, was clean and cheap, it was synthetic: In her womb was the future' (Koeppen 2002: 155–57).

Hazardous material: consumers redefine consumer citizenship

By the late 1960s, the optimistic vision of collective prosperity was curbed by the scenario of a 'landscape soon overflowing in every nook and cranny with litter and rubbish'. Non-biodegradable synthetics were seen as responsible for this trend. In his 1971 book *Müllplanet Erde* [*The Earth – A Planet of Waste*], Hans Reimer declared: 'Fighting waste amounts to either fighting the excrescences of plastic consumption or fighting specific characteristics of plastics' (Reimer 1971: 62; on the West German history of plastic waste, see Westermann, 2013). The piling up of plastic waste seemed to undermine the idea of a steady flow of goods as a mechanism of social integration. In 1973, the suspected carcinogenicity of the vinyl chloride monomer (VCM), from which polyvinyl chloride was made, gave rise to new concerns about the impact of plastic. The concerns were substantiated soon afterwards. Severe forms of liver cancer (hemangiosarcoma) in workers were found to relate to the production process of vinyl (Westermann 2007: 239–88). The discussion of plastic waste was an expression of the crisis the West Germans were facing due to the inadvertent consequences of their consumption-based political culture. The news of occupational cancer in the vinyl industry only aggravated the situation. Recognition of the hitherto unacknowledged material forces of plastic prompted consumers to reclaim their political power and speak out against vinyl consumption. By turning vinyl into an object of political concern, people added a new facet to vinyl's communicative function. Consumer citizens initiated a change in political culture, and, while doing so, challenged the alleged passivity of the consumer citizen fostered by the early conceptions of West Germany's consumer democracy.

The hot phase of anti-plastic activism started in the small city of Troisdorf, near Cologne. Once VCM's carcinogenicity was confirmed, a lobby group filed a suit on behalf of 40 out of the 130 workers directly involved with the production of PVC at Dynamit Nobel AG in Troisdorf. Taking the hazards of thalidomide as an example, the lobby group wanted the government and Dynamit Nobel AG to establish a foundation that would take care of the victims, both financially and medically. The West German VCM and PVC industry comprised 6,500 workers in the early 1970s, 1,000 of whom were directly exposed to the vinyl chloride monomer. Counted from the time of its mass industrial inception in the late 1930s, around 3,600 workers had presumably been exposed to the monomer. In May 1974, 80 cases of occupational health disease were reported, with most cases found among the employees of Dynamit Nobel AG.

Local activism continued. The health hazards were neither Dynamit Nobel's in-house problems nor singular cases to be dealt with by occupational health and safety administration alone. Activists shifted the boundaries of the workplace beyond the factory to the city council, by pointing out that the workers of Dynamit Nobel were part of Troisdorf's citizenry:

> You might say this is not our concern. Health and safety hazards at the workplace are none of the city's business. We approach the issue differently, since in our view, a city has an all-encompassing responsibility towards its citizens, including their workplace conditions.[18]

In their view, the VCM health hazards added to Troisdorf's long-standing conflict with the company. Complaints about Dynamit Nobel had been made regularly since the factory had extended its production facilities in 1966. For the immediate neighbourhood, the circumstances of production had proved to be scandalous. Noise and phenol pollution had been a constant problem, and the risk of explosions threatened adjacent houses.[19]

In Troisdorf, widening the local group of those concerned turned occupational hazards into broader environmental problems for the city. The same strategy of generalization was applied on a national level. However, exposing the VCM-related health hazards on a national platform did not depend so much on the toxic situation in and near the PVC production plants as it did on the plastic itself. News about the carcinogenicity of the VCM, released in late 1973, caused the national press to look into the consequences for everyone: 'Will we have to renounce plastics?' (n.a. 1974). As with plastic waste, the public considered the dramatic success of vinyl, its omnipresence, to have become its most problematic feature. Journalists and critics typically approached the issue by enumerating the most diverse applications of vinyl: 'Whoever talks of plastics, has to talk of vinyl. Almost everything can be made out of it, for example plastic buckets, floor coverings or bags' (n.a. 1975); 'Tubes and shutters, chamber pots and daily requisites, curtains and films, trinkets and trash, nearly everything can be made out of vinyl' (n.a. 1973). A citizens' initiative, addressing parliament, feared that 'the cancer risk extends to the *whole population* because PVC products like bottles, cans, boxes etc., are widely used. Safety factors and protection measures should be far-reaching enough to encompass vinyl's entire consumership'.[20]

The threat of cancer, latently existent in older critiques of synthetics and their 'chemicalizing' of the modern everyday, became firmly attached to vinyl and, by implication, other thermoplastics (Westermann 2007: 293–301). Accumulating and metabolized in the human body, the chemical agent VCM did not need any particular sociological framing in order to become apparent as a biological force. Vinyl also developed social force because the plastic's properties and toxic presence made people reconsider the very foundations of consumer democracy. Citizens put forward their critique in terms of highest values or the 'common good' (for a very useful sociology of critique, see Boltanski and Thévenot 1991). To the common goods of shared prosperity and freedom of choice, as defined by the social market economy, they added the value of physical integrity. Letter writers insisted that health was the most important common good that the federal state must guarantee. Drawing on paragraph 2 of the Federal Republic's Constitutional Law, they defended and reclaimed their right to health. Fearing that this right was being trampled on, a 'Working Committee for the Environmental Protection of East Frisia' wrote in response to the TV documentary *PVC – a Danger and its Downplaying*, 'What will you do to ensure that the consumer's fundamental right to personal physical integrity is protected against the risks of dealing with materials containing vinyl?'[21] In another line of argument, consumers wanted to be given information about dealing with plastics. They contended that the lack of plastic product information undermined the idea of electoral freedom. In their opinion, only those who were well informed would be able to exercise the consumer citizen's right to choose and thus be the all-decisive market force envisioned by the social market economists. Ministry officials dealing with consumer protection acknowledged that, as a rule, 'the average consumer' lacked technical information about science-based products like thermoplastics, and thus agreed with the analysis of many consumers. As one official put it, while generalizing the problem created by plastics: 'It seems to me, that this is the actual weak point in our social market economy.'[22]

Conclusion

Does vinyl have material politics? The question recalls Langdon Winner's essay of 1980, 'Do Artifacts Have Politics?' in which he argues that artefacts – in particular, large technical systems – are made and established to settle political issues in a community. Describing how technologies work towards historical change or to preserve continuity, Winner's essay has become a blueprint for understanding how political processes are embodied in materials and artefacts. Once put in place, artefacts transform both the built environment and the social order according to certain political preferences. Due to their materiality, artefacts and infrastructure tend to cement the power relations that led to their implementation. Winner advocates a soft version of technological determinism. The material form and built-in logics of a certain technology shape organizational logics, as well as the thinking and routines of individual users.

Materials are artefacts of a special kind. I have suggested in this chapter that the social impact of materials might best be captured by studying their properties and changing meanings attached to these properties. Vinyl makes for a striking example of soft technological determinism because it embodies a deliberately engineered flexibility. Vinyl made people interpret both its versatility and the chemical causes of its versatility in discursive ways, by simple use or by way of engineering and design. In a pragmatic approach, I have explored people's dealings with vinyl because the material politics of vinyl are not self-imposing. I have asked whether and how we can show that some explicit or implicit ideas of social order are at work in using or not using this plastic.

In a pragmatic approach, I have explored people's dealings with vinyl because the material politics of vinyl are not self-evident. I have asked whether and how we can show that some explicit or implicit ideas of social order are at work in using, not using or prohibiting this plastic. Materials do not simply vanish once they have 'congealed into objects' but 'continue to mingle and react' (Ingold 2007, 9). Not only do they threaten 'the things they comprise with dissolution', as Ingold warns. They can also pose a threat to consumers of these materials due to hazardous emissions or form a serious challenge to waste management due to their toxicity or non-biodegradability. Such forces may only unfold over time – which is why historians need to engage in charting the flows of plastic and many other materials.

The material politics of vinyl have proven powerful throughout the twentieth century. Vinyl was indispensable in bringing about West Germany's consumer democracy, but vinyl's material properties became implicated in the political legitimation of very different political regimes. Whether in National Socialist Germany or during the regime of the German Democratic Republic, twentieth-century German governments delivered or began to deliver their promise of abundance by filling shelves, furnishing homes and designing public spaces with vinyl and other plastics (Stokes 2000a).

Notes

1 The material and arguments presented here are based on my book *Plastik und politische Kultur in Westdeutschland* (2007); as for vinyl in the National Socialist economy of war, the chapter re-evaluates the material gathered in the book.
2 The changes in West German political culture were not brought about by vinyl alone. See Westermann, 2013.
3 Hoechst Company Archives UA BASF D 03.4/1 Kuko Wissenschaftliche Referate Bitterfeld, Düneberg, Eilenburg etc, 2. Sitzung, Referat Dr. Klatte-Rheinfelden, Über Vinylchlorid, no date, 1 and 6.
4 The road had been taken earlier: see Petzina (1968: 250) for the 'fuel contract' of 1933.
5 Federal Archives BA Freiburg RW 19/2484, GEHEIM, An Wa A, Sicherstellung des Rohstoffbedarfs für Kunstwerkstoffe, 1936.
6 BA Berlin R 3112/172, fig. 5 Produktion der thermoplastischen Kunststoffe 1936–42, 16.4.1940.
7 BA Berlin R8 VIII/245 Akten betr. Erzeugung, Rohstoffzuteilung und Verteilung von Kunststoffen, besonders von Polyvinylchlorid, Igelit, Weichmacher und Mipolam.
8 BA Berlin R 8 VIII/238 Deutsche Linoleumwerke AG Werk Delmenhorst, 20.11.44, betr. Igelit: 1.
9 UA BASF D 03.4/1 Kuko Wissenschaftliche Referate Bitterfeld, Düneberg, Eilenburg etc., 12. Sitzung, Referat Pungs, Troisdorf, 8.6.1934, Praktische Aussichtsmöglichkeiten für die neu entwickelten Kunststoffe: 4–5.
10 BA Berlin R 3112/172, Aufgaben und Ziele der weiteren Entwicklung der Kunststoffe, Prof. Kramer 25.4.40.
11 '"Schaffendes Volk". Große Reichsausstellung in Düsseldorf' 1937: 200. The 'Plastics Exhibit Hall' was covered with vinyl flooring, see Kaufman 1969: 173.
12 Trials of War Criminals 1953: 279. In Paris, the US company B.F. Goodrich also featured vinyl as a material of the future; see Blackford and Kerr 1996: 182.
13 State Archives LA Berlin Sign. 06930, Berlin plant. Ein Erster Bericht. Professor Hans Scharoun, in: Ausblick. Aufbaunachrichten der 'Berliner Ausstellungen'. Berlin im Aufbau. Schau der Arbeit und Planung für das NEUE BERLIN, 1946: 7.
14 LA Berlin Sign. 06930, Berlin plant: 7.
15 BA Koblenz B 102/9301, Vermerk 23.9.1952 Lehrschau, Anhang: Die Lehrschau Kunststoffe auf der Fachmesse und Leistungsschau Deutsche Kunststoffe v. 11. bis 19.10.1952: 1.
16 BA Koblenz B 106/25191, VKE Frankfurt, 12.1.1971 an BMI Hösel: 3; B 106/12853, Statement RG Verpackung für die Anhörung am 13.10.77 vom 10.10.1977, 1; BA Koblenz B 106/25131 Niederschrift über die Besprechung am 2. Juli 1969 im BM für Gesundheitswesen in Bad Godesberg, statement of the chemical industry: 4.
17 BA Koblenz B 106/25131 Niederschrift über die Besprechung am 2. Juli 1969 im BM für Gesundheitswesen in Bad Godesberg, statement of the chemical industry, Anlage 1. According to an industry newsletter quoted in Schmidt-Bachem (2001: 240, note 1), every West German used 20 plastic bags in 1971. On the history of plastic packaging from an environmental and cultural history perspective, see Nast (1997); Teuteberg (1995); Hawkins (2010).
18 BA Koblenz B 149/27871, Dr. Wilhelm Nöbel am 21.1.1974 im Rat der Stadt.
19 City Archives Troisdorf W.I.G. 2.3.II Dynamit Nobel AG 1972–74, Gerhard E. an Minister Deneke, 27.12.1973.
20 BA Koblenz B 149/27872, Hannah S., Gesellschaft für menschliche Lebensordnung e. V. Leer 18.10.1976.
21 BA Koblenz B 149/27872 Arbeitskreis für Umweltschutz Norden/Ostfriesland, 14.10.1976.
22 BA Koblenz B 102/9293a, Dr. Schaller 5.11.59, Kunststoffe, hier: Unterrichtung der Verbraucher über Kunststoffe und deren Eigenschaften.

References

Abelshauser, W. (1998) 'Germany: Guns, Butter, and Economic Miracles', in M. Harrison (ed.) *The Economics of World War II: Six Great Powers in International Comparison*, Cambridge: Cambridge University Press, 122–76.

——(2004) *German Industry and Global Enterprise: BASF, the History of a Company*, Cambridge: Cambridge University Press.

Betts, P. (2001) 'The *Nierentisch* Nemesis: Organic Design as West German Pop Culture', *German History* 19(2): 185–217.

Birkenfeld, W. (1964) *Der synthetische Treibstoff 1933–1945: Ein Beitrag zur nationalsozialistischen Wirtschaft – und Rüstungspolitik*, Göttingen: Musterschmidt.

Blackford, M. and Kerr, K.A. (1996) *BFG Goodrich: Tradition and Transformation 1870–1995*, Columbus, OH: Ohio State University Press.

Boltanski, L. and Thévenot, L. (1991) *De la justification: les économies de la grandeur*. Paris: Gallimard.

Böttcher, K. (1946) 'Von der Retorte zum Kunststoff-Montagehaus', *Neue Bauwelt* 1(13): 3–7.

Carter, E. (1997) *How German is She? Postwar West German Reconstruction and the Consuming Woman*, Ann Arbor, MI: University of Michigan Press.

Diefendorf, J. (1993) *In the Wake of War: The Reconstruction of German Cities after World War II*, New York: Oxford University Press.

Dommann, M. (2009) '"Be Wise – Palletize": die Transformationen eines Transportbrettes zwischen den USA und Europa im Zeitalter der Logistik', *Traverse* 16(3): 21–35.

Eckell, J. (1937) 'Schaffendes Volk', *Der Vierjahresplan* 1: 5.

Ehlich, K. (ed.) (1989) 'Über den Faschismus sprechen – Analyse und Diskurs', in K. Ehlich (ed.) *Sprache im Faschismus*, Frankfurt: Suhrkamp, 7–34.

Enzensberger, H.M. (1962 [1960]) 'Das Plebiszit der Verbraucher', in *Einzelheiten*, Frankfurt: Suhrkamp.

Erhard, L. (1957) *Wohlstand für alle*, Düsseldorf: Econ.

Evans, R. (2011) 'The Myth of the Fourth Reich', *New Statesman*, 24 November, www.newstatesman.com/europe/2011/11/germany-european-economic (accessed 20 June 2012).

Francoeur, E. (1997) 'The Forgotten Tool: The Design and Use of Molecular Models', *Social Studies of Science* 27(1): 7–40.

Freyer, H. (1957) 'Das soziale Ganze und die Freiheit des Einzelnen unter den Bedingungen des industriellen Zeitalters', *Historische Zeitschrift* 183: 97–115.

Friedel, R. (1983) *Pioneer Plastic: The Making and Selling of Celluloid*, Madison, WI: University of Wisconsin Press.

Furukawa, Y. (1998) *Inventing Polymer Science: Staudinger, Carothers, and the Emergence of Macromolecular Chemistry*, University Park, PA: University of Pennsylvania Press.

Girschik, K. (2010) *Als die Kassen lesen lernten. Eine Technik-und Unternehmensgeschichte des Schweizer Einzelhandels, 1950 bis 1975*, München: Beck.

Grosser, D. *et al.* (1996) *Deutsche Geschichte in Quellen und Darstellungen Bd. 11: Bundesrepublik und DDR 1969–1990*, Stuttgart: Reclam.

Hausen, J. (1947) 'Häuser aus Kunststoff?', *Kunststoffe* 37(1): 9–10.

Hawkins, G. (2010) 'Plastic Materialities', in S.J. Whatmore and B. Braun (eds) *Political Matter: Technoscience, Democracy, and Public Life*, Minneapolis, MN: University of Minnesota Press, 119–38.

Hayes, P. (1987) *Industry and Ideology: IG Farben in the Nazi Era*, London: Cambridge University Press.

Hilton, M. (2007) 'Consumers and the State since the Second World War', *Annals of the American Academy of Political and Social Science* 611: 66–81.

Hirschman, A.O. (1982) *Shifting Involvements: Private Interest and Public Action*, Princeton, NJ: Princeton University Press.

Hoechst Company Archives (1966) *Gründung der wissenschaftlichen Kunststoffkommission (Kuko) im Jahre 1930*, Dokumente aus Hoechst-Archiven: Beiträge zur Geschichte der Chemie 16, n.p.

Ingold, T. (2007) 'Materials against Materiality', *Archaelogical Dialogues* 14(1), 1–16 and 31–38.

Jarausch, K. (2005) 'Amerikanische Einflüsse und deutsche Einsichten. Kulturelle Aspekte der Demokratisierung Westdeutschlands', in A. Bauerkämper, K. Jarausch and M. Payk (eds) *Demokratiewunder. Transatlantische Mittler und die kulturelle Öffnung Westdeutschlands 1945–1970*, Göttingen: Vandenhoeck&Ruprecht, 57–84.

Kaufman, M. (1969) *The History of PVC: The Chemistry and Industrial Production of Polyvinyl Chloride*, London: Maclaren.

Koeppen, W. (2002 [1953]) *The Hothouse*, New York: W.W. Norton.

Kogon, E. (1964 [1949]) 'Deutschland von heute', in *Die unvollendete Erneuerung. Deutschland im Kräftefeld 1945–1963*, Frankfurt a. M: Europäische Verlagsanstalt, 102–120.

Koller, T. (1893) *Die Surrogate. Ihre Darstellungen im Kleinen und deren fabrikmäßige Erzeugung. Ein Handbuch der Herstellung der künstlichen Ersatzstoffe für den praktischen Gebrauch von Industriellen und Technikern*, Frankfurt a.M.: Bechthold.

König, R. (1965a [1956]) 'Masse und Vermassung', in *Soziologische Orientierungen. Vorträge und Aufsätze*, Köln: Kiepenheuer&Witsch, 479–83.

——(1965b [1959]) 'Gestaltungsprobleme der Massengesellschaft', in *Soziologische Orientierungen. Vorträge und Aufsätze*, Köln: Kiepenheuer&Witsch, 461–78.

König, W. (2004) *Volkswagen, Volksempfänger, Volksgemeinschaft. 'Volksprodukte' im Dritten Reich: Vom Scheitern einer nationalsozialistischen Konsumgesellschaft*, Paderborn: Schöningh.

Krohn, W. (1989) 'Die Verschiedenheit der Technik und die Einheit der Techniksoziologie', in P. Weingart (ed.) *Technik als sozialer Prozess*, Frankfurt: Suhrkamp, 15–43.

Latour, B. (1983) 'Give Me a Laboratory and I will Raise the World', in K. Knorr-Cetina and M. Mulkay (eds) *Science Observed: Perspectives on the Social Study of Science*, London/Los Angeles: Sage, 141–70.

Leysieffer, G. (1935) 'Kunststoffe aus deutschen Rohmaterialien', *Chemiker-Zeitung* I: 6–8.

Lindner, S. (2008) *Inside IG Farben: Hoechst during the Third Reich*, London: Cambridge University Press.

Löffler, B. (2002) *Soziale Marktwirtschaft und administrative Praxis: Das Bundeswirtschaftsministerium unter Ludwig Erhard*, Stuttgart: Steiner.

Lorentz, B. and Erker, P. (2003) *Chemie und Politik: Die Geschichte der Chemischen Werke Hüls 1938–1979*, München: Beck.

Meikle, J.L. (1995) *American Plastic: A Cultural History*, New Brunswick, NH: Transaction.

Merritt, A.J. and Merritt, R.L. (eds) (1970) *Public Opinion in Occupied Germany: The OMGUS Surveys 1945–49*, Champaign, IL: University of Illinois Press.

Müller-Armack, A. (1956) 'Soziale Marktwirtschaft', in *Handwörterbuch der Sozialwissenschaften* Bd. 9, Stuttgart: Gustav Fischer und Mohr, 390–92.

n.a. (1937a) 'Die Kunststoff-Industrie auf der Reichsausstellung "Schaffendes Volk" Düsseldorf', *Kunststoffe* 27(6): 159–62.
n.a. (1937b) 'Kunststofftagung des Fachausschusses für Kunst-und Pressstoffe Düsseldorf 12. und 13. Mai', *Kunststoffe* 27(6): 162–63.
n.a. (1937c) '"Schaffendes Volk": Große Reichsausstellung in Düsseldorf', *Plastische Massen in Wissenschaft und Technik* 7: 200.
n.a. (1952) 'Verlauf der Kunststoff-Tagung und Kunststoff-Messe in Düsseldorf', *Kunststoffe* 42(12): 468–75.
n.a. (1973) 'Gefährlicher Kunststoff', *Der Spiegel* Nr. 50: 147.
n.a. (1974) 'Krebsverdacht bestätigt', *Die Zeit* 27: 37.
n.a. (1975) 'Das unheimliche Vinylchlorid', *Süddeutsche Zeitung* 30 April–1 May: 13.
Nast, M. (1997) *Die stummen Verkäufer. Lebensmittelverpackungen im Zeitalter der Konsumgesellschaft. Umwelthistorische Untersuchung über die Entwicklung der Warenpackung und den Wandel der Einkaufsgewohnheiten*, Berlin, Bern: Lang.
Petzina, D. (1968) *Autarkiepolitik im Dritten Reich*, Stuttgart: Deutsche Verlags-Anstalt.
Plumpe, G. (1990) *Die IG-Farbenindustrie-AG: Wirtschaft, Technik und Politik 1904–1945*, Berlin: Duncker & Humblot.
Reimer, H. (1971) *Müllplanet Erde*, Hamburg: Hoffmann und Campe.
Reuben, B.G. and Burstall, M.L. (1973) *The Chemical Economy: A Guide to the Technology and Economics of the Chemical Industry*, London: Longman.
Saechtling, H. (1952) 'Welche Aufgaben verfolgt die Lehrschau bei der Fachmesse Kunststoffe 1952?', *Kunststoffe* 42(7): 203.
Schelsky, H. (1965 [1956]) 'Gesellschaftlicher Wandel', in *Auf der Suche nach Wirklichkeit. Gesammelte Aufsätze*, Düsseldorf: Diederichs, 337–51.
Schildt, A. (2000) 'Die 60er Jahre in der Bundesrepublik', in A. Schildt, D. Siegfried and C.-C. Lammers (eds) *Dynamische Zeiten. Die 60er Jahre in den beiden deutschen Gesellschaften*, Hamburg: Hans Christians, 21–53.
Schmidt-Bachem, Heinz (2001) *Tüten, Beutel, Tragetaschen: Zur Geschichte der Papier, Pappe und Folien verarbeitenden Industrie in Deutschland*, Münster: Waxmann.
Schwippert, H. (1951) 'Das Bonner Bundeshaus', *Neue Bauwelt* 6(17): 65–72.
Shove, E., Watson, M., Hand, M. and Ingram, J. (2007) *The Design of Everyday Life*, Oxford: Berg.
Spoerer, M. (2005) 'Demontage eines Mythos? Zu der Kontroverse über das nationalsozialistische "Wirtschaftswunder"', *Geschichte und Gesellschaft* 31: 415–38.
Stoeckhert, K. (1950) *Kunststoffe ohne Geheimnis. Einführung in ihr Wesen, ihre Verarbeitung und ihre Anwendung*, Kevelaer: Butzon&Bercker.
——(1975) 'Der Markt für Kunststoff-Verpackungen in der BRD', *Kunststoffe* 65(4): 241.
Stokes, R. (2000a) 'Plastics and the New Society: The German Democratic Republic in the 1950s and 1960s', in S. Reid and D. Crowley (eds) *Style and Socialism: Modernity and Material Culture in Post-war Eastern Europe*, New York: Oxford University Press, 65–80.
——(2000b) 'Privileged Applications: Research and Development at I.G. Farben During the National Socialist Period', in D. Kaufmann (ed.) *Geschichte der Kaiser-Wilhelm-Gesellschaft im Nationalsozialismus: Bestandsaufnahme und Perspektiven der Forschung*, Göttingen: Wallstein, 398–410.
Stuart, H.A. (1952) *Die Struktur des freien Moleküls: Allgemeine physikalische Methoden zur Bestimmung der Struktur von Molekülen und ihre wichtigsten Ergebnisse*, Berlin, Göttingen, Heidelberg: Springer.

Teuteberg, H.-J. (1995) ‘Die Rationalisierung der Warenpackung durch das Eindringen der Kunststoffe und die Folgen’, *Environmental History Newsletter*, Special Issue 2: 112–48.

Thévenot, L. (2001) ‘Pragmatic Regimes Governing the Engagement with the World’, in T.R. Schatzki, K. Knorr Cetina and E.V. Savigny (eds) *The Practice Turn in Contemporary Theory*, London: Routledge, 56–73.

Tolischius, O.D. (1937) ‘Reich Shows Uses of its Materials’, *The New York Times*, 9 May: 6.

Tooze, A. (2006) *The Wages of Destruction: The Making and Breaking of the Nazi Economy*, London: Allen Lane.

Treue, W. (1955) ‘Hitlers Denkschrift zum Vierjahresplan’, *Vierteljahrshefte für Zeitgeschichte* 3: 204–10.

Trials of War Criminals (1953) *Trials of War Criminals Before the Nuremberg Military Tribunals under Control Council Law No. 10, Nuremberg, October 1946–April 1949*, Washington, DC: US Government Printing Office.

Tschanter, E. (1946) ‘Die Kunststoffhäuser der Ausstellung “Berlin plant”’, *Kunststoffe* 36(2): 34.

Westermann, A. (2007) *Plastik und politische Kultur in Westdeutschland*, Zürich: Chronos.

——(2008) ‘Die Oberflächlichkeit der Massenkultur. Plastik und die Verbraucherdemokratisierung der Bundesrepublik’, *Historische Anthropologie* 16(1): 8–30.

——(2013) ‘When Consumer Citizens Spoke Up: West Germany’s Early Dealings with Plastic Waste’, *Contemporary European History* 22(3).

Wiesen, J. (2011) *Creating the Nazi Marketplace: Commerce and Consumption in the Third Reich*, Cambridge: Cambridge University Press.

Winner, L. (1980) ‘Do Artifacts have Politics?’ *Daedalus* 109: 121–36.

5 Paying with plastic

The enduring presence of the credit card

Joe Deville

Reflecting on VISA's early years, its founder, Dee Hock, draws attention away from the material composition of the credit card and towards what he sees as its real function:

> The [credit] card was no more than a device bearing symbols for the exchange of monetary value. That it took the form of a piece of plastic was nothing but an accident of time and circumstance. We were really in the business of the exchange of monetary value.
>
> (Hock 2005: 143)

Hock's marginalization of the role of plastic is to be expected, given the way in which monetary objects are thought about – both popularly and in social scientific discourse – as things that come to matter because of their ability to negate their materiality and act as passive mediators of value. This chapter, however, opens up for examination this simple 'accident of time and circumstance', and explores how the plastic in the plastic card can indeed matter to the composition of the consumer credit market assemblage, in both moments of borrowing and default.[1]

All monetary objects – credit cards as much as bank notes and coins – are material things that can be considered as 'economic devices' or 'economic *agencements*' (Muniesa *et al.* 2007; Mcfall 2009).[2] At its simplest, this means that a monetary object – like many objects and processes – 'renders things, behaviors and processes economic' (Muniesa *et al.* 2007: 3).[3] They do not do so alone, however. The success of the particular object inserted between the economic exchange of an ideal typical buyer and seller is dependent on the security and predictability of its relationship with a diverse and complex range of associated actors, both human and non-human.

Yet it is not simply the transactional device's insertion into a broader network of processes that matters.[4] For, as the recent material turn in economic sociology has shown, and drawing on a longer interrogation of materiality in science and technology studies (STS), devices themselves can shape the course of economic action.[5] Martha Poon's work provides a relevant example: she shows how the credit score, as it has moved into the related but distinct arena of mortgage underwriting, is a device that has had specific impacts on the conditions of economic possibility, generating 'novel pathways of micro-economic market participation which have gradually become amplified … into macroeconomic circuits of capital flow' (Poon 2009: 669).

The objects at the centre of this chapter do not need to be brought into view from behind the scenes of consumer credit; the cards that smooth so many people's passages through consumer spaces are a recognizable part of our everyday lives. These are the devices that are at the 'frontline' of many forms of consumer credit-enabled economic exchange (not all – for instance, unsecured bank loans are also a form of consumer credit). Returning to Michel Callon's classic account of economic 'framing' (Callon 1998: 199), they are objects that enable the stabilization of transactional frames, allowing an ideal typical buyer and seller to enter an economic space and, via the handover and acceptance of the card, be 'detached', able to leave each other without any unwanted connections or 'attachments' (also described in terms of movements of 'entanglement' and 'disentanglement') (Callon 1998; Callon *et al.* 2002; Callon and Muniesa 2005; Muniesa *et al.* 2007; Mcfall 2009; Muniesa 2009).[6]

Such moments of stabilization are also necessarily provisional, inherently involving concomitant processes of exclusion and externalization. Processes of detachment are thus always dependent on the 'outside' of the moment of exchange which can, depending on circumstance, re-intrude. Viviana Zelizer (1997, 2002) has shown how this can apply to monetary objects as much as any other, tracking the way they can be imbued with social understandings and, as a corollary, become implicated in generating social consequences that are quite unrelated to the economic value associated with them. It is thus not only the value of monetary media that matters, but also the socially derived 'values' that come to be attached to them.[7]

Yet Zelizer, like Hock, misses the fact that the material composition of the monetary medium might *also* be important in shaping social and economic life. Specifically, hers is an understanding in which the social acts through, or becomes attached to, forms of money, with the materiality of money assumed to be the *a priori* condition for the expression of cultural values. This is not a general critique: in many situations, the materiality of the monetary object may *not* obviously matter, for the very reason that powerful, performative socio-technical assemblies exist to render their material properties often of little relevance to everyday users. However, as this chapter argues, this is – depending on the actor or historical moment – not always either relevant or successful or desired. In these cases, the sociomaterial composition of monies may indeed matter. This is thus an argument for considering monetary objects as continually processual entities, the material presence of which should never be assumed to be an irrelevance. That is to say, monetary objects always retain the capacity, under certain conditions, to become directly implicated in causal effects that are generated co-constitutively with their users.

The remainder of the chapter is divided into two sections. The first focuses on the United States, given its status as the historical test bed for consumer credit, the experimentation of which paved the way for technologies that would subsequently be adopted and adapted worldwide. It opens up the way in which plastic and its specific sociomaterial affordances have, in different and changing ways, mattered in the history of the payment card and in creditors' engagements with users. The initial focus is on how the material presence of the payment card emerged as a problem for card issuers and manufacturers looking to expand the scope of the consumer credit market, and on the role different materials played in solving this problem. It then proceeds to examine how the mass market issue of the plastic card itself played a key role in accelerating the first major credit card boom. The second part moves to the United Kingdom, exploring the role of the card not in moments of borrowing, but at times of default – that is, when consumer credit borrowers, for whatever reason, are unable (or unwilling) to meet their obligations to repay.

A history of the early credit card: the changing monetary affordances of plastic

The origins of the mid- to late twentieth-century consumer credit boom can be traced back far beyond the introduction of the credit card (see Calder 1999; Hyman 2011). However, as this section will show, focusing on a period running from the early 1950s until around 1970, it was the plastic card around which earlier developing tendencies crystallized. The card not only responded to the needs of a market that was rapidly expanding, but also played an active role in enlisting a vast untapped reserve of American borrowers. The credit card and the consumer credit market therefore need to be seen as very much co-emergent, with the former solving many of the problems posed by the latter.

At the same time, it is worth noting that, after an initial period of experimentation and innovation, credit cards quickly became relatively stable monetary objects. Especially in their form, credit cards now still very much resemble credit cards then; by contrast, the sophisticated behind-the-scenes technologies of credit scoring and profiling have become almost unrecognizable from their rudimentary antecedents. In this sense, the formal endurance and stability of this particular object needs to be explained. Is this simply a case of material 'lock-in'? Or is there something about this particular sociomaterial solution to the problems posed by the credit market that is as relevant now as it was then?

The continuity between past and present is made clear by looking at one of the first plastic credit cards, issued by American Express in 1959 (see Figure 5.1).[8] A number of features are instantly recognizable: the card bears a distinctive printed design; it also includes, as its principal security feature,

Figure 5.1 Example of the first plastic American Express credit card
Source: (Image used courtesy of American Express Company)

space for a signature; and finally, the card also displays key identifying information, including name, expiry date and a unique numerical ID. This number is printed in a machine-readable font, to enable semi-automated processing of the sales invoices onto which the number would be transferred via an imprinting device. However, why plastic? Here we have to go back a little further, into the pre-history of credit cards, to explore what plastic, as a particular monetary material, offered that other materials did not.

Early American consumer credit was merchant or industry specific, initially centred around individual department stores but soon being extended to a network of merchants owned by oil companies and telegram providers (see Calder 1999; Hyman 2011; Marron 2009; Stearns 2011). From a relatively early stage, the card (in many cases, a term that accurately described the material) – or a similar wallet-sized object (I will return to this) – emerged as essential to these forms of consumer credit. These objects, bearing the name of the merchant alongside details of the account holder, enabled the carrier 'to identify one's account to a centralized credit system that was available at multiple locations, or multiple merchants within an industry-specific network' (Stearns 2011: 6). As with many forms of early US consumer credit, these were 'charge cards', which meant that they enabled the extension of credit on a limited basis, with the account expected to be repaid in full at the end of the month.

However, one problem with these cards was that information had to be hand-copied from the card to the sales draft, inevitably leading to errors. An early solution was the 'Charga-plate' system, which was introduced into select

Boston department stores in 1928 and continued to play an important role in America's credit market well into the 1950s (Hyman 2011). It combined a system of embossed, roughly credit card-sized metal plates with an accompanying 'imprinter', which would transfer the customer's details, via carbon paper, on to the sales draft (Stearns 2011: 8–9). Quite aside from the additional robustness of these metal devices, their key benefit was that the information they contained was more *reliably transferable*. For example, in a pilot project run by Standard Oil of California in 1952 using a later version of these metal fobs, errors of identification were cut by 94 per cent, while the time spent writing out sales slips was cut in half (Mandell 1990: 23). These metal cards were thus able to carry individualized information in (effectively) three dimensions – and, crucially, this information became transferable when inserted into an accompanying technological infrastructure. In this context, the particular problem that the metal used in these cards moved towards solving was – in a very Latourian sense – a problem of translation.

Early experiments with plastic attempted something similar: an ultimately unsuccessful example was the Simplan system. It combined a wallet-sized plastic card, perforated with small rectangular holes (effectively a unique customer number), with the customer's name and address embossed onto the card. The card would be inserted into a reader that would transfer this embossed and punched information on to the resulting invoice, tying it to the customer and producing a carbon copy. This was then processed, part mechanically, at a district processing office (Gregory 1955: 57–58; *National Petroleum News* 1955). The accompanying marketing literature promises 'prompt billing', 'fewer errors in transcription' and a 'less costly' process, as well as indicating a customer preference for this new material, with '[c]ustomer surveys show[ing] that the one-piece plastic card will be preferred over check-book or metal plate styles of customer's name and account number' (*National Petroleum News* 1955). The Simplan system, aided by plastic, thus claimed (but, in not securing widespread adoption, ultimately failed) further to remedy the translation problems that surrounded consumer credit-based market exchange at the time.[9] Later plastic cards, such as the embossed American Express card in Figure 5.1, can be seen as refinements of the ambition expressed here. American Express also similarly drew on the promise of progress and modernity that seemed to accompany the spread of these new plastic devices: 'Soon … a *plastic* card', proclaimed a heading in one of its advertisements (*The New York Times* 1958: emphasis in original).

However, the success of the plastic card did not just lie in its ability to transfer information more accurately, or in the potential associations plastic had with a modern, forward-moving age. While for much of the 1950s metal fobs were the principal material for the semi-automated passage of consumer credit information, it has to be recognized that for most of the decade it was (by comparison) an unsophisticated, far cheaper and far lighter cardboard card that transmitted the bulk of American consumer credit transactions. For it was Diners Club – enabling travellers to enjoy the convenience of being able to pay at participating restaurants without the need for cash – and not the merchant-centred metal cards that provided the business model for the later boom in consumer credit.

While also a charge card, to be repaid in full monthly, Diners Club was different because, while being tied to a type of business, it was not tied to any *one* business. This changed the market dynamics of credit considerably. Suddenly, consumer credit became focused not on the extension of credit to existing customers around a circumscribed set of locations, but rather on the attraction of *new* customers around an autonomous *credit product*, in which it was not the specific locations that mattered but rather the size of both a customer base and a merchant network. What Diners Club realized was that, in order to build the latter, you needed the former: a large customer base made it more attractive for restaurants to take the card; more participating restaurants, in turn, made Diners Club more attractive for potential cardholders.[10] This hunt for users translated into the hitherto unprecedented conversion of mass marketing tactics into the world of consumer credit. Notably, this included direct mail tactics to attempt to convince potential users to make applications, helping membership climb from around 150,000 in 1952 to around 1 million by 1960 (Simmons 1995: 34–37).

Yet, while a trailblazer, Diners Club did not become the model for the credit card as we recognize it today. It failed in part because it did not combine this distribution model with a card that could offer revolving credit – repayable, with interest, over a potentially indefinite period – as later credit cards did. Diners Club also failed to recognize the promise of automation that plastic opened up. As Matty Simmons – then head of sales marketing at Diners Club – admits, reflecting on the later success of American Express, '[t]he plastic card, like computerized billing, was something we all knew would come, but it had to come with imprinters, and Amex simply moved more quickly with all these new facets of the business than did Diners Club' (Simmons 1995: 83). What Simmons doesn't mention – and neither Diners Club nor American Express understood – is that the plastic card was itself a device that could be used to boost customer numbers. In the history of the payment card, it is thus not American Express but Bank of America, in the development and launch of its BankAmericard (a precursor of VISA), that provides the more significant historical model for the later expansion of the consumer credit industry.

The plastic BankAmericard was launched more or less at the same time as American Express's first plastic card. While the two cards were formally very similar, Bank of America realized that plastic offered an additional affordance that would go on to be exploited again and again in the following decade. With the combination of The Databosser – a high-speed embossing machine able to read customer information fed in via IBM punch-cards – and a new type of plastic that fitted into these embossing machines previously used to produce metal cards, Bank of America had access to the most efficient, most cost-effective way of mass producing payment cards at the time (Dashew and Klus 2010: 159–69).[11]

This efficiency and ability to produce lightweight cards in volume was to prove crucial, as Bank of America demonstrated in the experiments surrounding the launch of the BankAmericard in California in September 1958 (e.g. see Guseva 2005; Nocera 1994; Wolters 2000). Bank of America selected the city of Fresno, population 130,000, for a trial of a new marketing strategy, following ideas stemming from in-house innovator Joe Williams. After convincing a base of over 300 businesses to take part, and accompanied by an advertising blitz, they mailed a plastic BankAmericard, unsolicited, to every Bank of America customer in the city. As Joseph Nocera writes, this 'drop' was Joe Williams's solution to the chicken and egg of problem of how to create both customers *and* participating merchants (Nocera 1994: 26). In this respect, it worked. In the months following, another 800 retailers in the area joined the BankAmericard scheme (Wolters 2000: 332). The direct mailing strategy was then rolled out statewide, in part to pre-empt similar tactics from competitors, with 1959 seeing Bank of America placing one of the largest orders for printing machines and credit cards in the history of the industry (Dashew and Klus 2010: 168). By the end of the year, the bank's annual report was proudly proclaiming that 2 million of its customers were holding BankAmericards (Wolters 2000: 333), all made of plastic and all embossed with individual customer information.[12] For the first time, the core elements were in place for a new consumer credit market model: this involved a ruthless focus on increasing a customer base by doing more or less whatever it took to get into a customer's hands a card that gave access to revolving credit, a card that could also be inserted into a semi-automated payment processing system, and a card that could be manufactured and individualized quickly and dispatched cheaply *en masse*.

It took some years for this model to be perfected, but, once it was, it provided the basis for an approach that would define the period of rapid expansion of the US credit market in the late 1960s. It was only *after* the United States had been saturated with credit cards that the next important set of innovations – the widespread adoption of customer-profiling technologies such as credit scores – began to shape the consumer credit industry.[13] Before this could occur, the sociomaterial payment infrastructure and customer/merchant base had to be in place. The consequence was that in a period from around 1966 to 1970 – until the practice was eventually made illegal – around 100 million plastic credit cards were mailed, unsolicited, to potential users, in what the president of the Bank of America, Rudolph Peterson, called 'the great credit-card race' (Nichols 1967; see also O'Neil 1970; Nocera 1994: 54; Stearns 2011: 33). One contemporary report, drawing on an American Bankers Association survey, suggests that, at least in the earlier years, only around half of these new credit card offerings were accompanied by credit checks (Furness 1968: 65). Some events in the period became infamous.

In Chicago in 1967, for example, a series of competing regional banks mailed 5 million cards to the city's residents, with some families in the more attractive suburbs claiming to have received up to 15 cards, including – if reports are to be believed – toddlers and convicted criminals (Jordan 1967; Nocera 1994: 54).

There is, however, more to this unsolicited mass mailing than just securing market share. This can be illustrated via another credit card experiment, this time presented to a 1968 government hearing on the practice by a spokesperson from the American Bankers Association. In the experiment, New York-based Marine Midland Bank sent 33,357 promotional credit card application forms to potential users. They received responses from only 221, or 0.7 per cent. However, when cards were sent directly to 731 recipients, 19 per cent were actively using them within 60 to 90 days (Bailey 1968: 24). The spokesperson expanded on this theme:

> At the time of the receipt of the application [the recipient] may not have any use for it, and at that point doesn't think he wants to have a card. However, if he has a card in his pocket, he knows his credit has been established … I am sure he will welcome the opportunity to use it when the time comes.
>
> (Bailey 1968: 27)

Having arrived, as another witness put it, 'like a gift from heaven' (Jackson 1968: 32), physically present, ready to be used as and when required, with no application needed and no need to assess at the moment of receipt whether or not it would be needed in future, the card was not just slightly more attractive to borrowers, it was significantly so. Indeed, the data from this experiment were provided as unashamed evidence of the very need for the practice of unsolicited credit card mailings, being seen as an essential part of the armoury for banks seeking to challenge the competition. However, at the hearing this effect was also seized upon by critics, who provided numerous case studies of individuals who, by virtue of having received an unsolicited card, had been encouraged into borrowing at levels they would not, it was asserted, have entered into otherwise.

Something similar is also captured in a *Life* magazine article from 1970 on 'the great plastic rush' (O'Neil 1970: 55). The article is accompanied by a large double-page cartoon (Figure 5.2) showing the US population being physically smothered by an avalanche of plastic cards, fired on them from a giant canon by banks, with bankers on computers and in their offices trying to understand and control, after the fact, 'the new plastic society' they have unleashed.[14] The piece imagines a typical conversion of a person (here presumed to be male even though women were also targeted) from a 'sorehead', actively annoyed by and resistant to the solicitations of the creditor, to a borrower:

Figure 5.2 Understanding the new plastic society: 'In a rush to "get their plastic on the air," banks randomly fired off credit cards. Computers – key to controlling them – are still trying to catch up'

Source: (O Neil 1970: 48–49; illustration by John Huehnergarth)

> The average man, his senses dulled by an endless reception of junk mail, simply chucks his card into a desk or cupboard or dresser drawer if he is among those who are not instantly galvanized by their bank's sudden new interest in their well-being. He may eventually betray reactions characteristic of the sorehead group: when the bank send him a follow-up statement … he may poke holes in it or punch it full of staples and send it back to confuse the bank's computer. But as long as the card stays in his dresser he is subject, though he is not aware of it, to a curious and sort of subconscious temptation. Bank records indicate he will eventually dig it out and give it a try and will thereafter tend to use it again … and again …
>
> (O'Neil 1970: 50)

Both the experimental and anecdotal evidence points towards these new plastic credit cards acting as 'lures for feeling' (Whitehead 1978: 88; see also Fraser 2009; Halewood 2003; Stengers 2008). By virtue of their very material presence in the homes and lives of so many Americans, they exerted a pull on them – a nagging draw towards the possibilities that credit, with its ability to shift economic value forward in time, promises. They thus became, to return to the terms outlined at the start of this chapter, indispensable devices for market attachment. These cards were, and continue to be, lively entities of their own, reaching out to the user as the user might reach out to them. As we now know, this small, mutually constitutive reaching out, between hand and card, happened on countless occasions, providing part of the fuel for the increasingly global, diverse credit explosion the consequences of which were arguably only beginning to be fully realized at the beginning of the twenty-first century.

It is to this quite different historical moment that this chapter now moves. This also means moving from a moment in which political debate on credit turns around questions of the expansion of lending, to one in which it turns around the expansion of default. Drawing in part on conversations and letter evidence posted on the most prominent British online consumer discussion forum, it examines how the card can play a role not only in the proliferation of credit borrowing, but also in processes designed to retrieve some of that borrowed money from those who are unable or unwilling to repay. As we will see, here the card becomes a site for contestations over power and ownership.

Cut up your card: the credit card becomes collections device

After decades of familiarization with credit cards, it is not surprising that their material composition does not bother those for whom these devices have become a stable, unremarkable component of their movement through spaces of consumption. They are, for many, simply 'there', in purses, pockets and wallets, alongside a range of other objects. However, the payment card can assume an explicitly agential role for those users of transactional forms of consumer credit who find themselves unable to repay their debts.

Here, it is important to understand that consumer credit payment cards are monetary objects that are delivered accompanied by explicitly laid-out conditions of use, usually outlined in a credit agreement. These range from detailing how the card's attachment to the user should be secured – by being signed and the PIN memorized – to limits as to the card's transferability (e.g. 'by the named individual alone'). There is one particular condition designed to follow both user and card around: the continued ownership of the card by the creditor. This is a residual legal claim that, in some cases, is made explicit on the card itself, or in a credit agreement. This retention of the card's ownership appears to be in part connected to the demands that the creditor makes on the borrower's conduct, of the kind just outlined. Yet there are instances where the ownership of the card can matter in a quite different way. For credit agreements can also contain the provision to revoke the user's right to the card and to demand its destruction. One situation in which this can come into play is in the management of default.

Consumer debt collection, undertaken by the creditor or an external collector working on their behalf, involves the routine deployment of letters and phone calls, in ways that seek to generate the debtor's calculative attention. In the specific case of letter design, collectors use a range of strategies to grab the attention of the defaulting debtor. One of these, used by some in the earlier stages of default, centres around the payment card itself. Take the following extracts from the collections letters of major UK creditors:[15]

> All credit cards are the property of American Express Services Limited and we prohibit their further use. The cards must be cut up and immediately returned to us at our above address.
>
> (American Express)

> If you have not already returned or destroyed your card(s), we request that you do so immediately, cut in half for security reasons.
>
> (HSBC)

In terms of the journey of these creditors' plastic cards alongside the respective borrowers, these letters are therefore envisaged as denoting an end point. On one hand, such demands need to be seen as one of the many detachment devices that are integral to the collections process – and to the consumer credit market more widely. The user who was previously extended both a borrowing facility and an obligation to repay becomes reconstituted more firmly in relation to their obligations, while the creditor seeks to prevent the debt escalating further. On the other, demanding that the card be cut up and returned might seem somewhat peculiar, given: (i) the often large sums of money owed by defaulting debtors to creditors; (ii) the (by comparison) relatively tiny value of the card device; (iii) the usual willingness of creditors to replace lost or damaged cards at no cost; and (iv) the ability of contemporary card issuers to block electronically most unauthorized uses of contemporary payment cards.

Among defaulting debtors themselves, it is a demand that is often read with some suspicion and/or confusion, as the following posts on the Consumer Action Group online debt advice forum show:

> … as well as a separate letter asking me to cut up the card and send it back to them along with any cheques! They know very well that even if I wanted to use the infamous card it wouldn't be accepted so what's the point?[16]
>
> If a credit card company writes and asks you to cut up the card and return it, can it be assumed that they have terminated the account?[17]
>
> the letter they sent to me asks me to cut up and return my card back to them, but when i called the manager [he] told me to keep hold of it, which do you think i am best to do?[18]
>
> i did call them as soon as i [received] the letter telling me to return my credit card cut up … cos i thought to myself 'what the bloody hell they playin at?'[19]

The first poster expresses incredulity that they are being asked to return a card that is now technologically incapacitated. The second wonders whether the demand marks the passing of a particular threshold. The third suggests a discrepancy within the lending organization itself as to whether it in fact wants the card back at all.

You could respond to these queries by exploring the risks that continually circulating cards – electronically blocked or not – still pose to the creditor, or whether or not the blocking of the credit card does in fact mark the passing of a definitive threshold (it does not). However, I suggest that the third and fourth posts point towards perhaps the most valuable function served by a creditor's demand for 'their' credit card: that the *questions* generated by this demand might prompt the borrower to contact the creditor – as, in fact, both these borrowers did (even if, as in the third post, the question in fact remains unresolved).

More particularly, collections communications frequently play with the opacities that characterize many of our journeys through modern social life. These include the difficulty of understanding the language and functioning of legal processes – as is the case here. Unless the reader has the expertise to do the work of reading *against* the demand by the creditor for their cards, their choices are to live with this indeterminacy, seek recourse to outside expertise – as these posters are trying to do – or to contact the creditor.

By asserting its continuing claim over the device – demanding back redundant, effectively worthless pieces of plastic and electronics, cut into pieces – the creditor attaches its collections activities to the affordances of the payment card. The creditor draws attention to the device that, more than any other, stands as the material embodiment of the relationship between the creditor and the borrower. By demanding its destruction and return, the creditor hopes that it can both symbolically mark the end of the relationship as it was previously constituted, and prompt the borrower into remedial action. Here, by drawing attention to its materiality – and hence its fragility – the collector transforms this plastic borrowing device into a small but potentially potent collections device.

On this particular forum, the responses by other contributors to such prompts are manifold, but there is one theme of particular interest, which focuses attention not just on the way in which 'meaning' may be attached to these objects but also on how their material composition can matter. This can be summed up by an image posted by one debtor, Davey, in response to a demand for their card made by American Express. It shows a close-up of a thumb and forefinger holding an American Express card, which is being held over a candle flame, which is melting one corner of the card. Above the card is the caption, 'Let them ask for the card back now', with 'Get stuffed Amex!' scrawled on to the card in marker pen.[20] This user echoes the tone of others on the site who variously imagine shredding cards into fine strips before sending them back,[21] or talk gleefully about cutting cards into tiny pieces and burying them in the garden.[22] Or there are those who respond warmly to Davey's picture posted on the forum:

> Quality … we should make a huge credit card bonfire and film it.
>
> (Talbot)

> I have two nice platinum colored cards to add (they've sent me a nice brand spanking new [one] even though I've told them [I'm] in financial difficulty … nothing like responsible lending [smiley face icon].
>
> (the_shadow)

Here, in a refrain with echoes of those that were heard in America in the late 1960s, the_shadow claims to have been sent cards – understood as enticements to borrow – against his wishes. However, now these cards become – or are imagined as becoming – the target of a form of micro-political protest. Both the perceived absence of any value inherent in the apparently worthless plastic object and the very possibility of their easy destruction – as the creditor's initial demand indeed points towards – allows the card to become a vector for a very material retribution enacted on the otherwise largely dematerial figure of the creditor. The creditor, meanwhile, is held to be culpable not simply for providing a borrowing facility, but also for providing the borrowing *device*.

Conclusion

An archaeologist unearthing a pre-modern coin on a dig will have no trouble placing it in a direct analogical relationship to the ostensibly similar-looking coins in their pocket. Here, the idea of money demonstrates a certain endurance over time, even as the engagements between the user and monetary object vary.[23] However, as Isabelle Stengers writes, 'endurance is never an attribute, always an achievement: throughout its adventures, something … succeeds in maintaining some thread of conformity between past and present' (Stengers 2005: 44). Stengers is drawing attention to the way in which entities of all sorts always exist in dialogue with the processes and entities around them. This includes the very 'idea' of money itself, which – despite its apparent stability – is an abstraction with no inherent guarantee of historical continuity. It includes the processual materialities of money's transactional objects. For the unearthed coin's second achievement is to have stubbornly and materially endured.

In the case of plastic payment cards, this chapter has opened up some of the reasons behind their emergence and endurance from a period starting in the late 1950s to the present. It has also shown how their own processual materiality can also, in important and changing ways, matter. In the 1950s and 1960s, the development of new forms of plastic cards first promised, then offered, a solution to the problem of how to deliver, in a cost-effective way, a lightweight, individualized, three-dimensional information carrier into the hands of borrowers which, when inserted into the accompanying socio-technical infrastructure, would convey this information faster and more accurately across time and space. In part, it is thus the reliable reproducibility of the consumer credit transaction that becomes one of plastic's achievements, providing the crucial foundations for the subsequent rapid expansion and stabilization of the consumer credit market.

At the same time, these new plastic cards were to become enrolled as vital actors in the expansion of the US consumer credit customer base. Like many other plastic objects, the combination of a material that was cheap, robust and just the right kind of malleable meant that millions of cards could be sent out to US citizens, with the cost of production and distribution insignificant in light of the potential profit and market share at stake. Again, like other plastic objects, these unsolicited cards were clearly coded as disposable; they *could* be thrown away by the recipient. However, the issuer was not only hoping, but, banking on the fact, that in many cases they would not. The expectation was that, after their arrival through the letterbox, these millions of cards would materially endure in US citizens' drawers and homes, until a moment came when their promise of being able to move future value into the present became relevant enough to prompt their use.

In the case of consumer credit default, this routine material co-presence with the borrower offers a different opportunity. Here, the card becomes an affordance not for securing new market attachments, but for re-securing the economic value associated with an existing attachment between consumer and creditor. It does so by seeking to draw attention to the relations of obligation and ownership implied by the user's use of the particular credit product, with the card coming to stand as the material embodiment of this otherwise dematerialized market relationship. One unintended effect, however, is to provide debtors with an ostensibly fragile, destructible object upon which they can enact small but very physical retributions against the creditor. In these minute protests, the socioeconomic politics surrounding the proliferation of consumer credit borrowing are, for a brief instant, brought to bear on the usually unnoticed credit card itself. In so doing, they direct attention towards what usually goes unremarked: that the accumulation of billions of dollars of consumer credit debt not only depends on, but also has been stimulated by, the accumulation in so many everyday lives of millions of superficially unremarkable plastic cards.

Notes

1 Credit cards are now not simply plastic, but ever more sophisticated plastic-electronic composites. Yet it should be recognized that these objects are popularly understood *as* plastic, as the title of this chapter and a number of popular and academic books suggest. For instance, see Brown and Plache (2006); Essig (2010); Evans and Schmalensee (1999); Rowlingson and Kempson (1994).

2 See also Fabian Muniesa's account of money as an 'attachment device', also drawing on Zelizer (Muniesa 2009).

3 The distinction between 'economic' and 'market' here refers to the often-neglected observation that not all economic processes are necessarily market processes. This

is particularly relevant given the long history of monetary objects being used in a range of settings.

4 See Çalışkan and Callon's (2009) detailed exploration of the opportunities and limits of, for instance, the 'embeddedness' approach characteristic of much so-called 'new economic sociology'.

5 See, for instance, Noortje Marres and Javier Lezaun's (2011) exposition of the history of the device, considered as an object involved in enacting forms of material participation.

6 On the potential complexity and variability of what factors are brought inside particular transactional frames, see Daniel Miller's (2002) critique of Callon.

7 See David Stark (2009) on the long-standing separation of research into questions of value (the domain of economics) from that into questions of values (the domain of economic sociology).

8 Credit card issue date confirmed by correspondence with American Express. Oil companies issued the very first plastic charge cards in the mid-1950s (Simmons 1995: 91; see also Marron 2009: 81).

9 The degree of sociomaterial investment that is still going into trying to solve this problem can be illustrated by reference to the central place of testing in card manufacturing industry (Turner 2009; Rolfe 2010).

10 Drawing on Simmons (1995: 15–38) and Grossman (1987: 261–77).

11 American Express at the time was using a system operated by a competitor of Dashew Business Machines, presumed to be the slower Addressograph-Multigraph embosser. American Express, however, switched to Dashew machines in 1961 (Dashew and Klus 2010).

12 There is not scope here to go into the details of the problems that Bank of America subsequently ran into with their particular strategy; this has been extensively detailed elsewhere (Guseva 2005; Nocera 1994: 328–33; Wolters 2000: 333). However, as both Guseva (2005) and Wolters (2000) argue, in the long run the strategy proved a commercial success for the bank.

13 A process documented in relation to the case of Fair, Isaac and Company, creators of the FICO score, by Poon (2007).

14 Many thanks to John Huehnergarth for allowing this image to be used. I am also grateful to Nils Huehnergarth for his assistance.

15 These are drawn from letters uploaded by defaulters onto the Consumer Action Group 'Debt Collection Industry' sub-forum to be discussed shortly; for the sake of brevity, only two examples are shown.

16 www.consumeractiongroup.co.uk/forum/showthread.php?253822-flowerchild-V-MBNA-PPI-Charges-reclaim.

17 www.consumeractiongroup.co.uk/forum/showthread.php?214861-Cut-up-card.

18 www.consumeractiongroup.co.uk/forum/showthread.php?136848-Termination-of-Egg-credit-card-agreement&p=1927121& viewfull = 1.

19 www.consumeractiongroup.co.uk/forum/showthread.php?153426-hsbc-have-cancelled-my-credit-card-without-any-warning.

20 www.consumeractiongroup.co.uk/forum/showthread.php?184309-Davey-vs-Amex.

21 www.consumeractiongroup.co.uk/forum/showthread.php?179236-Tragic-case-of-Judgement-because-a-cut-up-card-was-produced-by-MBNA-at-hearing.&p=1938126&viewfull=1#post1938126.

22 www.consumeractiongroup.co.uk/forum/showthread.php?179236-Tragic-case-of-Judgement-because-a-cut-up-card-was-produced-by-MBNA-at-hearing.&p=2027669&viewfull=1#post2027669.

23 This argument has a history: notably, it was Georg Simmel (2004) who argued that money, as an abstract ideal, sits in a dialectical relationship to its somewhat paradoxical, necessarily incomplete, empirical realization (see also Gilbert 2005: 362–63).

References

Bailey, T.L. (1968) 'Statement of Thomas L. Bailey on behalf of the American Bankers Association', in US Senate Committee on Banking and Currency (ed.) *Bank Credit-card and Check-credit Plans: Hearings before the Subcommittee on Financial Institutions of the Committee on Banking and Currency, United States Senate, Ninetieth Congress, Second Session, on Credit Cards, October 9 and 10, 1968*, Washington, DC: US Government Printing Office.

Brown, T. and Plache, L. (2006) 'Paying with Plastic: Maybe not so Crazy', *University of Chicago Law Review* 73(1): 63–86.

Calder, L. (1999) *Financing the American Dream: A Cultural History of Consumer Credit*, Princeton, NJ: Princeton University Press.

Çalışkan, K. and Callon, M. (2009) 'Economization, part 1: Shifting Attention from the Economy Towards Processes of Economization', *Economy and Society* 38(3): 369–98.

Callon, M. (1998) 'Introduction: The Embeddedness of Economic Markets in Economics', in M. Callon (ed.) *The Laws of the Markets*, Oxford: Blackwell.

Callon, M., Méadel, C. and Rabeharisoa, V. (2002) 'The Economy of Qualities', *Economy and Society* 31(2): 194–217.

Callon, M. and Muniesa, F. (2005) 'Peripheral Vision: Economic Markets as Calculative Collective Devices', *Organization Studies* 26(8): 1229–50.

Dashew, S.A. and Klus, J.S. (2010) *You Can Do It! Inspiration and Lessons from an Inventor, Entrepreneur, and Sailor*, Los Angeles, CA: Constellation Press.

Essig, L. (2010) *American Plastic: Boob Jobs, Credit Cards, and Our Quest for Perfection*, Boston: Beacon Press.

Evans, D.S. and Schmalensee, R. (1999) *Paying with Plastic: The Digital Revolution in Buying and Borrowing*, Cambridge, MA: MIT Press.

Fraser, M. (2009) 'Experiencing Sociology', *European Journal of Social Theory* 12(1): 63–81.

Furness, B. (1968) 'Statement of Miss Betty Furness, Special Assistant to the President for Consumer Affairs, accompanied by Leslie V. Dix, Director for Legislative Affairs', in US Senate Committee on Banking and Currency (ed.) *Bank Credit-card and Check-credit Plans: Hearings before the Subcommittee on Financial Institutions of the Committee on Banking and Currency, United States Senate, Ninetieth Congress, Second Session, on Credit Cards. October 9 and 10, 1968*, Washington, DC: US Government Printing Office.

Gilbert, E. (2005) 'Common Cents: Situating Money in Time and Place', *Economy and Society* 34(3): 357–88.

Gregory, R.H. (1955) 'Document Processing', in *Papers and Discussions Presented at the November 7–9, 1955, Eastern Joint AIEE-IRE Computer Conference: Computers in Business and Industrial Systems*, New York: ACM, doi.acm.org/10.1145/1455319.1455328 (accessed 16 April 2012).

Grossman, P.Z. (1987) *American Express: The Unofficial History of the People who Built the Great Financial Empire*, New York: Crown.

Guseva, A. (2005) 'Building New Markets: A Comparison of the Russian and American Credit Card Markets', *Socio-Economic Review* 3(3): 437–66.

Halewood, M. (2003) 'Subjectivity and Matter in the Work of A.N. Whitehead and Gilles Deleuze: Developing a Non-essentialist Ontology for Social Theory', unpublished thesis, Goldsmiths, University of London.

Hock, D. (2005) *One from Many: VISA and the Rise of the Chaordic Organization*, San Francisco: Berrett-Koehler.

Hyman, L. (2011) *Debtor Nation: The History of America in Red Ink*, Princeton, NJ: Princeton University Press.

Jackson, R.E. (1968) 'Statement of Royal E. Jackson, Chief of the Bankruptcy Division, Administrative Office, US Courts', in US Senate Committee on Banking and Currency (ed.) *Bank Credit-card and Check-credit Plans: Hearings before the Subcommittee on Financial Institutions of the Committee on Banking and Currency, United States Senate, Ninetieth Congress, Second Session, on Credit Cards. October 9 and 10, 1968*, Washington, DC: US Government Printing Office.

Jordan, D.D. (1967) 'Curbs on Credit Cards Issued by Banks being Sought by Patman in House Bill', *Wall Street Journal*, 29 August: 28.

Mandell, L. (1990) *The Credit Card Industry: A History*, Boston: Twayne.

Marres, N. and Lezaun, J. (2011) 'Materials and Devices of the Public: An Introduction', *Economy and Society* 40(4): 489–509.

Marron, D. (2009) *Consumer Credit in the United States: A Sociological Perspective from the 19th Century to the Present*, New York: Palgrave.

Maurer, B. (2006) 'The Anthropology of Money', *Annual Review of Anthropology* 35: 15–63.

Mcfall, L. (2009) 'Devices and Desires: How Useful is the "New" New Economic Sociology for Understanding Market Attachment?', *Sociology Compass* 3(2): 267–82.

Miller, D. (2002) 'Turning Callon the Right Way Up', *Economy and Society* 31(2): 218–33.

Muniesa, F. (2009) 'Attachment and Detachment in the Economy', in P. Redman (ed.) *Attachment: Sociology and Social Worlds*, Manchester: Manchester University Press.

Muniesa, F., Millo, Y. and Callon, M. (2007) 'An Introduction to Market Devices', *Sociological Review* 55(2): 1–12.

National Petroleum News (1955) 'Sohio Launches IBM Credit Card System', 47: 81.

The New York Times (1958) 'When You Have an American Express Credit Card', 13 November: 25.

Nichols, R.E. (1967) 'B of A's Peterson Cautions Banks on "Credit-card Race"', *Los Angeles Times*, 17 March: C14.

Nocera, J. (1994) *A Piece of the Action: How the Middle Class Joined the Money Class*, New York: Simon & Schuster.

O'Neil, P. (1970) 'A Little Gift from your Friendly Banker', *Life*, March: 48–58.

Poon, M. (2007) 'Scorecards as Devices for Consumer Credit: The Case of Fair, Isaac & Company Incorporated', in M. Callon, Y. Millo and F. Muniesa (eds) *Market Devices*, Oxford: Blackwell, 284–306.

——(2009) 'From New Deal Institutions to Capital Markets: Commercial Consumer Risk Scores and the Making of Subprime Mortgage Finance', *Accounting, Organizations and Society* 34: 654–74.

Rolfe, A. (2010) 'Welcome', *Payments, Cards & Mobile*, Nov./Dec. (Test Tools Supplement): 3.

Rowlingson, K. and Kempson, E. (1994) *Paying with Plastic: A Study of Credit Card Debt*, London: Policy Studies Institute.

Simmel, G. (2004 [1900]) *The Philosophy of Money*, London: Routledge.

Simmons, M. (1995) *The Credit Card Catastrophe: The 20th Century Phenomenon that Changed the World*, New York: Barricade Books.

Stark, D. (2009) *The Sense of Dissonance: Accounts of Worth in Economic Life*, Princeton, NJ: Princeton University Press.

Stearns, D.L. (2011) *Electronic Value Exchange: Origins of the VISA Electronic Payment System*, New York: Springer.

Stengers, I. (2005) 'Whitehead's Account of the Sixth Day', *Configurations* 13: 35–55.

——(2008) 'A Constructivist Reading of *Process and Reality*', *Theory, Culture & Society* 25(4): 91.

Turner, A. (2009) 'Welcome', *Payments, Cards & Mobile*, Mar./Apr. (Test Tools Supplement): 3.

Whitehead, A.N. (1978) *Process and Reality*, New York: The Free Press.

Wolters, T. (2000) '"Carry Your Credit in Your Pocket": The Early History of the Credit Card at Bank of America and Chase Manhattan', *Enterprise and Society* 1(2): 315–54.

Zelizer, V. (1997) *The Social Meaning of Money: Pin Money, Paychecks, Poor Relief, and Other Currencies*, Princeton, NJ: Princeton University Press.

——(2002) 'Intimate Transactions', in M.F. Guillén, R. Collins, P. England and M. Meyer (eds) *The New Economic Sociology: Developments in an Emerging Field*, New York: Russell Sage Foundation, 274–300.

Part III

Plastic bodies

6 The death and life of plastic surfaces

Mobile phones

Tom Fisher

This chapter discusses moments when our engagements with plastic objects 'touch' us and elicit feelings that influence the actions we take with them. Noting that these feelings seem to be associated with the quality of plastic surfaces as they age and wear, it focuses on touch-screen mobile phones and their accoutrements. These new manifestations of plastic surfaces draw on well-embedded narratives about materials and at the same time introduce new inflections on our embodied relationship with them. These new actions are motivated both by efforts to protect the surfaces of phones from damage, and by what the phone is – a device that 'extends' our bodies through its function in communication. They are generated both by economic rationales – phones are expensive – as well as the affective dimension of the practices of which phones are part (Reckwitz 2002; Shove and Pantzar 2005). They include a repertoire of small habitual body movements – such as rubbing the phone to keep it clean – as well as decisions to use other plastic devices, like screen protectors and cases, to preserve them.

To attend to the feelings just mentioned, the chapter develops the idea that changes in the surfaces of plastics that accompany their ageing produce feelings of disgust in us (Fisher 2004). While these feelings are relatively mild, they can make us pay attention to the plastic surfaces of our possessions, and may lead us to rid ourselves of them. It proposes that the ageing plastics of touch-screen phones and their accoutrements may stimulate embodied disquiet because of the relatively close physical contact we have with them. In exploring this relationship, I introduce the results of preliminary interview and survey work with touch-screen phone owners. There is a strong relationship between feelings of disgust and evidence of *our* death (Falk 1994: 86; Rozin 1999: 24), and a link is proposed between this embodied emotion and our efforts to forestall the 'death' of plastic objects; we mark our relationship with these objects at a visceral level, and this affects our actions.

Plastics' death is implicit in the relationships between plastic objects' physical properties and the cultural narratives that have grown up around them in the 150 years since they first appeared. Plastic operates simultaneously at both ends of various extremes – 'gross *and* hygienic', as Barthes (1987: 54) famously put it. Plastic objects can be ephemeral *and* too persistent – think of plastic carrier bags fluttering in the breeze. Their physical properties give them the propensity to attract us and to disappoint us, to acquire disquieting qualities and sometimes disgust us. Some objects are 'cool' because they are made of plastic; others are 'tacky' for the same reason. Plastic can connote utter technological advancement *and* barely adequate performance. The material can be hyper-palpable (in the various objects called 'rubbers') or pass us by entirely in everyday life (in paint, for instance). This chapter concentrates on the former, considering examples of 'palpable' plastics – things we can easily tell are plastic.[1]

All of these types of object are part of the personal environments that people manage in everyday life, part of our 'extended self' (Belk 2001). They are us extended into objects, elements of the material culture that forms us as social beings through the dialectical process that Daniel Miller (1987) terms 'objectification'. The chapter concentrates on ways in which the materiality of plastic objects means they are more than matter to us – they matter to us; we mind about them – picking up on a discussion between Tim Ingold and Daniel Miller (2007) about the significance of what can be thought of as the physical 'essence' of materials. The extent and nature of our 'minding' about plastics is quite difficult to pin down *because* plastics are so ubiquitous, but it is possible to see it in the processes whereby plastic objects turn from being useful extensions of our selves into the world to being things we want to void from our spatial bodies, to get rid of. This process entails both physical changes in the objects – plastics surfaces degrade in distinctive ways – and consequent changes in our estimation of them. It is a process that is especially distinctive because plastics fall so far. They promise so much when new that when they have aged we are perhaps especially affronted by them, and in the process of this change their plastic-ness can become briefly illuminated by the attention we then pay to them.

It seems that the closer a plastic object comes to our body, the more acutely we evaluate it. Along with its potential to elicit visceral disquiet, plastic's characteristic way of degrading has effects that we reason through – it raises questions of hygiene. We often associate it with disposable, 'one-trip' items, which we accept for their convenience and sterility (Freinkel 2011: 120–21), but, although many plastic items have short lives that give us little opportunity to make strong relationships with them or much cause to reflect on them, many do not. We live with many plastic objects for some time, living with the processes that transform them from pristine to unacceptable. We are familiar with the characteristic softness and absorbency of plastic – it dirties, but is not clean-able. As an interviewee put it when talking about plastic food storage containers:

> [My partner] likes tomato soup. But it actually stains the plastic. It stains it … If it's going that way then it must be going the other way mustn't it, and it kind of bothers you a bit. It bothers me.
>
> (Fisher 2003: 162)

She indicates here that, in the process of ageing, plastics both acquire and give up matter – they absorb substances, so they may therefore exude dubious substances with potential consequences for health.

The chapter extends from this everyday learning about the relation between ageing plastics and our bodies using some initial findings from studying the ways in which people use sacrificial plastic surfaces to preserve their touch-screen phones. These are quite a new type of object, and their design requires a higher degree of physical interaction than previous types of phone: you have to touch them constantly, stroke them, caress them in order to use them. This has led the user-interface design community to re-evaluate the well-embedded hierarchy of the senses that prioritizes sight, to develop an understanding of the 'aesthetics of touch' (Motamedi 2007), retrieving a discussion that can be traced back to Herder's aesthetics (Benjamin 2011). Because phones are so salient to everyday practices of communication, display and adornment, the consequences of physical changes in their plastic surfaces that are appraised by both sight and touch are therefore both distinctive and relatively accessible – their owners are very ready to talk about them. Further, they have generated a variety of plastic accessories, cases and screen protectors that are intended to forestall the degradation of the phone by acting as relatively ephemeral sacrificial plastic surfaces.

The 'physicality' of plastics: cleaning, damage and texture

Plastic surfaces exhibit repertoires of degradation that we come to recognize as we live with plastic objects. There seem to be three ways in which they become degraded: by contamination with dirt; through mechanical damage; and by 'weathering' – stuff sticks to them, they get scuffed and they 'perish'. Although other materials share these processes of degradation, focusing on them tells us something about plastics' contradictory material nature. When the often peerless and perfect surfaces of plastics are penetrated by mechanical damage, we get to see that they are the same all the way through, like metal or ceramic, but softer. When stuff sticks to plastic, we can sometimes clean it off effectively (as long as we take care not to scratch the surface in the process), but often this is not possible. Plastics stain – polyethylene kitchenware takes on the colour of the tomato soup we store in it. Plastics also discolour as their structure is affected by light. As these processes take place, plastics both acquire dirt, and in distinctive ways themselves become 'matter out of place' (Douglas 1966).

In the process of caring for our 'spatial body' by physically engaging with the ways plastics degrade, we build our awareness of their structure. Grant McCracken (1988: 86) notes the degree to which the meaning of goods attaches to the goods themselves, and is sometimes 'supercharged' by the rituals that grow up round them and through which we maintain our relationships with them. He identifies maintenance among these rituals and we get to know about the properties of materials particularly by the attempts we make to forestall their degradation by cleaning them. There is advice available online about how to clean plastic items, which indicates that we need assistance in these maintenance rituals; it also illustrates the linkage between attempts to maintain plastic possessions, the understanding of their physical properties that results and the decision to get rid of them. One such website gives this guidance on cleaning melamine tableware:

> While cleaning melamine tableware is not tough, once it gets discoloured with frequent use and washing, it loses its appeal. Since scratches and burn marks cannot be removed from it, perhaps the best solution is to throw out the damaged pieces.
>
> (Banerjee 2002)

Clearly, cleaning plastics is not always possible. However, if it is important to an individual that they attempt to maintain their plastic possessions, they may seek techniques to clean them. The interviewee quoted above indicates that it is this type of direct everyday experience of plastics that provides the most compelling personal knowledge of their qualities – their relative hardness, softness and porosity – as we see them degrade. Through this direct bodily experience, we know we *can* clean glass, ceramic and bare metal effectively, because they are hard and 'fused' – we can use detergents and mild abrasives and the surfaces will stand up to some scrubbing. Other porous materials generate their own maintenance repertoires that may involve a sacrificial layer of (non-palpable) plastic. We know we can revive the interior of our houses by repainting them, using a plastic 'sacrificial' layer. We apply the same principle to wood and leather. Bare wood absorbs dirt that we can't get off without cleaning it so vigorously as to remove the surface of the wood – though the resultant damage can be valuable as 'patina' (Green 2006: 33). Wooden surfaces are usually sealed with a sacrificial layer of plastic varnish that will soak in and then set.

Our attempts to clean plastic items are often problematic because there is no 'underneath' to the surface that can be 'restored'. The porous surface formed from the tool against which they were moulded is all there is – it can't be replaced by a surface beneath, and can only sometimes be protected with a sacrificial layer. Phones, car interiors, kitchenware and children's toys do not have the fused surfaces of ceramic, glass and metal, and cannot be revived by being painted or polished. If their surfaces are worn, they tend not to respond well to being washed because the texture that wear produces on their surface is very good at holding on to the dirt they accumulate. Plastics' subtle absorbency,[2] combined with the wear and scratches that plastic surfaces all too easily acquire, are tell-tale signs of their objective material reality.

Plastics can degrade in more spectacular ways, too. An early widespread use for cellulose nitrate plastics was for film stock, which becomes potentially explosive as its chemistry alters through time, as well as becoming brittle. Museum curators who have plastic objects in their collections tell horror stories about the ways in which plastics alter over time (Keneghan 2009). In his early work, the modernist sculptor Naum Gabo was fond of using the most advanced plastic materials, and some of these pieces have been completely lost to catastrophic degradation (Rankin 1988). Of course, newer plastics are more stable – so stable in fact that when they appear in our surroundings they persist longer than we would like. However, none of them is completely stable, and all the thermoplastics have the same porous microstructure. This is significant to us in our everyday interactions with plastic objects because it locks together the disgust that their dirty surfaces can elicit with concerns for cleanliness and the pollution of our bodies by the chemicals that thermoplastics can exude (Freinkel 2011).

Plastics' materiality: from peerless perfection to 'hard and ugly'

Our embodied understanding of plastics' degradation is invoked in our everyday judgements about plastic objects. The negative associations with the materials that result from that degradation persist in the face of the plastics industry's efforts to promote a range of positive qualities of plastics (Meikle 1995: 28): lightness, flexibility, 'plasticity', modernity.[3] These are relatively abstract attributes, which relate to the specifiability of plastics and are prominent in discourses close to design, production and purchase. More relevant to this discussion are the features of plastics that characterize the materials as we experience and physically interact with them over time, their colours and smells included. For Cecilia Fredriksson, it was the *smell* of plastic that encapsulated the excitement of modern consumption in her account of a Swedish low-cost department store (Fredriksson 1997). The smell summed up the 'new world of things' that the store represented: positive values attached to the smell of plastic. However, the physical reality of what is being smelled might not denote the promise of a modern future, but rather quite the reverse. In the case of thermoplastics like polyvinyl chloride (PVC), the smell comes from the volatile elements of the material evaporating from it;[4] a key result of this volatility is the death of the material as it becomes less 'plastic'. One of Fredriksson's interviewees remembers one of her first purchases: 'a plastic make-up bag, purple and blue. Through time it got hard and ugly' (Fredriksson 1997: 124–25). Once it was hard, ugly and no longer desirable, it is safe to assume that it was thrown away. The object died along with its material.

The characteristic way that death overtakes plastic objects is informed by a recent debate between Tim Ingold and Daniel Miller about the 'materiality' of materials. In 2007, Miller contributed to a dialogue about the relationship between the physical qualities of materials and the very concept of 'materiality'. He was responding to a paper in which Ingold proposed that the properties of stone as it interacts with the other elements of the environment – air, water, etc. – are as telling of its materiality as the abstract ideas we might have of it, its meaning. In his critique of Ingold, titled 'Stone Age or Plastic

Age?' Miller (2007) suggests that plastic artefacts are more likely to be objects of concern in contemporary life than stone ones, and he offers mobile phones as an obvious example. Criticizing Ingold for a primitivist appeal to essential material properties as the ground of human experience of the environment, while acknowledging the validity of his 'sensitivity to the flow of material qualities' (Miller 2007: 27), Miller proposes that we encounter technologies, not materials. These he calls 'bundles of material propensities' rather than 'natural bundles of innocent properties' (Miller 2007: 26). While this is a critique of Ingold's radically material-centred view, Miller keeps with the agenda of *material* culture studies as a corrective to an over-emphasis on the symbolic register that he sketched out in the early 1990s (e.g. see Miller 1994).

Quite naturally, the many studies of mobile phones that have appeared over the last decade figure them as a means of communication, and as symptoms of shifts in the way we communicate and socialize. However, the material objects with which we physically interact to do this communicating and socializing receive rather less attention than the communication that they enable. Their 'material propensities' remain obscure. As Webb puts it, 'the mobile becomes a portal and the networks become data pipes that enable the basic connectivity' (Webb 2010: 65). However, Webb acknowledges that, alongside its communication function, a mobile phone has some similarities to clothing and jewellery, being 'an object of beauty and fascination'. Miller has himself contributed to the literature on phones (Horst and Miller 2006), and, to take up Miller's challenge to Ingold, I want to bring together some of the strands of my discussion of plastic by thinking about mobile phones not as elements in a complex web of communication but as objects with which many of us physically interact very frequently, objects that we carry close to our bodies, which begin to show their age, and which die.

In this, I address Horst and Miller's (2006: 7) injunction to study 'not things or people, but processes', concentrating on processes of *physical* interaction with phones, in particular the way their materials become compromised and the steps people sometimes take to forestall this.[5] Like Webb's characterization of phones as jewellery, this goes against a tendency to promote the immaterial aspects of communication technologies. However, the relative dematerialization characteristic of the consumption of communication technologies does not mean that its material manifestations are any less important. Every website requires a computer or a smartphone through which to access it. We are not 'jacked into' the network, as William Gibson (1984) suggested we would be – we handle *things*, to get glimpses of what some argue to be a disembodied humanity. Focusing on things in this way counters the suggestion that new technologies imply a disembodied humanity (Moravec 1995, and quoted in Pepperell 2005: 30), instead acknowledging what Pepperell (2005) argues is an extended/distributed embodiment. This 'extensionist' view seems to acknowledge the relevance of both human and technological 'bodies' – as Pepperell puts it, 'technology is embodied humanity'.

Phones, for instance, not only have the theatrical, performative presence implied by a 'portal', but are also like telescopes or peepholes. We engage physically and materially with the technological body of the phone to extend ourselves into the immaterial spaces to which they provide access. All bodies die, and, as portals/telescopes/peepholes for communication, phones die from the inside out; their innards stop working. As *things*, they die for us from the outside in, as they acquire traces of our physical interaction with them and in that process they demonstrate the qualities characteristic of their materials. I concentrate here on newer touch-screen phones rather than those with buttons, because the more tactile nature of their interface makes the qualities of their material salient to the process of using them, as indicated by the attempts that people make to forestall their material death by putting them in a case and by using a clear plastic film to protect the screen. 'Screen protectors' mean that physical interaction with the material of the screen is substituted by interaction with a plastic surface that has the same function as a clear sacrificial layer as in other familiar uses – in packaging, for instance (Fisher and Shipton 2009).[6]

The salience of the materials of touch-screen phones is confirmed by the fact that, although it is generally difficult to get interview participants to talk about plastics in the abstract, and it is necessary to go to specialist groups to get direct accounts of experiences of the material (Fisher 2004), preliminary work on touch-screen phones suggests that their owners are relatively forthcoming about their materials. Interviewees can speak lucidly about them as objects, how they show signs of wear, how they degrade over time and the steps they may take to forestall this. The shiny, immaculate surface of a new phone is perhaps the strongest signifier of its new-ness, and from interviews[7] with touch-screen phone owners it seems that many take active steps to preserve these qualities by using screen protectors that are available for phones and other touch-screen devices. Alongside preserving an 'as new' surface, there is a clear functional rationale behind screen protectors: to preserve the screen so it is possible to see clearly what it shows.[8] This suggests that the screen protector protects the phone both as a 'portal', in Webb's terms, and as an object of 'beauty and fascination'. However, phones are the object of a particular sort of fascination, one that has a strong physical dimension, alongside their likely status as objects of contemplation and imagination (Campbell 1987), which derives from their being designed to be handled constantly. This sacrificial plastic layer shares the protective function of the cases and covers, but it is different in that it *becomes* the surface through which the user interacts with the phone, as well as protecting it. It is ephemeral, and normally invisible in use – a screen usually looks the same with or without a screen protector – which implies that the motivation for using them is to know, rather than to see, that the phone is protected.

In contrast, the rubbery texture of some cases, which is emphasized in the way they are retailed, clearly aligns with the tactility necessary to touch-screen phones. Cases are designed to be visible: some have considerable material ‘presence’, and their design clearly draws from the visual vocabularies associated with their materials – they may be made of leather, knitted fabric or of diverse formulations of plastic. Whereas screen protectors are barely visible, a case may completely cover the phone, making it impossible for others to easily appreciate what brand it is except when it is being used. When they are made of ‘palpable’ plastic, the formulations used for phone cases offer a playful repertoire of textures and patterns based in the material character of the cases and the sensual effects that are played out through them. This playfulness is emphasized in the way the cases are packaged and displayed in phone shops, sometimes with a hole in the packaging so prospective purchasers can sample the tactile experience on offer by feeling the surface of the case through the hole – a ploy borrowed from the design of packaging for children’s toys. Like screen protectors, cases are relatively disposable, and some plastic ones emphasize the mechanical protection they provide by using the rubberiness of silicone polymers[9] to suggest that the phone will keep working even if you drop it because the case has sacrificed itself to save the phone. It is surfaces like these that can be touched in the shop through the hole in the packaging, and which presumably offer sensorial experiences that substitute for the material of the phone itself when they are used, confirming by proxy the importance of the tactility of the phone’s materials.

Motivations for protection

Along with the desirable tactility of touch-screen phones that is indicated by the playful character of some of the cases and the spontaneous gestural character of the interfaces themselves come several negative consequences. Physical contact between body and phone leaves traces, as does contact between body and clothing, or spectacles, or cutlery. Whereas the latter are routinely cleaned,[10] this is not such a simple matter with a rather delicate electronic device. Everyday observation suggests that a new repertoire of bodily comportment (Michael 2006: 58) has grown up round touch-screen phones to cope with the risk of damage and contamination. Observations have included an individual putting their phone down on a paper napkin in a café, as well as frequent sidelong appraising glances at the screen that result in careful wiping, on sleeve, tissue or trouser leg. These protective strategies are to an extent delegated to phone cases and screen protectors, but *these* objects then themselves become subject to the same tendency to damage and contamination through use.

Various factors are likely to influence the adoption of cases and protectors, and a recent survey[11] suggested that motivations for their use are indeed rather complex. The survey showed a relationship between gender and the rationales given for cases and screen protectors, with more females than males basing their use of cases on ‘personalization’ or the appearance of the phone (though this might be misleading, given possible variation in readiness to admit to this motivation between males and females). For one interviewee, a phone shop employee, it was clear that using a case and screen protectors was driven by her need to maintain a good impression to customers by having a pristine phone. It would be reasonable to expect that wanting to keep the phone clean is a widespread motivation to adopt cases and screen protectors, and the survey did suggest that the rationale for having them may often be to protect them from damage and keep them 'as new'. This line of thinking has an economic basis, with a strong correlation evident between using a case and an expressed intention to sell the phone when it is replaced.

There was an indication of a positive relationship between the degree to which respondents were concerned about the cleanliness of their surroundings in everyday life and their adoption of screen protectors, but also suggestions of a negative relationship between such a concern and using a case. This may indicate that, while it helps give a sense of security against damage, a case compounds the problem of keeping the phone clean because it traps the matter of dubious origin that collects on the phone's surface in its nooks and crannies. It is this matter which the rubbing of a touch-screen phone on the sleeve is attempting to remove – what one interviewee[12] identified as 'ear-prints'. The special way that phones pick up unwanted matter, and its effect on our feelings about them, points towards qualities shared with other plastic objects that either come into intimate contact with the body or imply the body's boundaries. In both cases, there is a potential to invoke feelings of disgust (see Fisher 2004).

The same interviewee described her way of avoiding ear-prints when using her touch-screen phone. She holds it in the usual relationship to her head, but a little way away, to avoid touching it with her ear. This individual has a particularly strong commitment to preserving her devices 'as new', describing this as a trait evident in her care for her childhood toys, and distinguishing it from tidiness. Many of her habits of care for her possessions employ plastics as a sacrificial layer, including her use of screen protectors on all her touch-screen devices and plastic bags to keep dust off the computers that she no longer uses, but keeps in the attic. This individual's desire to preserve her possessions seems to determine many of her actions with them, and the way she preserves her touch-screen devices builds on her established routines. She suggested that her use of screen protectors and cases was not driven by a need for cleanliness, which might be activated by these devices' close relationship to her body – despite her avoidance of ear-prints, she has never used the screen cleaner that came with her iPad. Instead, she seems motivated by a wish to preserve the coherence of the *object's* body, to stop *it* being compromised. As she puts it, the screen protector is about 'protecting it from scratches, from having lots of dirty marks on it, this isn't the same as having marks on … the actual thing itself. This is fine, but if I hadn't got this on I'd be really quite worried about doing that.'

She admitted to taking her iPad out of its case sometimes, just to hold it and look at its pristine surfaces and form. Her concern to preserve her iPad, and all her similar possessions – phone, MP3 player, computer – by using plastic as a sacrificial layer is driven by a concern for *their* surfaces. She was not especially interested in the screen protectors she uses – they are 'some sort of plastic' – but spoke clearly about their function to protect another plastic surface 'from scratches, from having lots of dirty marks on the actual thing itself'. The screen protector is clearly an inconsequential sacrificial plastic layer that can somehow absorb or filter out the dubious evidence of her physical interaction with the surface, indexed in finger- and ear-prints: 'I don't mind the prints being there but I'd rather them be on the filter, on the protector rather than on the thing.'

This relatively intense concern to save the plastic surfaces that matter to her from signs of deterioration extends to the habits of others. She admitted to being bothered by her hairdresser keeping her 'naked' iPod Nano in her handbag. This interviewee clearly distinguishes between her relationship with the surfaces of her possessions and her relationship with the sacrificial plastic layer; one matters, the other does not. Speculatively, it can be suggested that to the extent that the physical contact with these surfaces denotes an 'extension' into them – a 'distributed embodiment' – then to protect those surfaces implies that there is a desire to take special care of that extended body, to postpone its death.

Another interviewee, a student user-interface designer, also gave clear justifications for her use of a case and screen protector on her phone. She spent £500 on the phone because she is designing for its interface and she has a screen protector on its glass front and keeps it in a rather beautiful white leather case, which obscures the surfaces of the sides and back. She rationalized this in terms of its monetary value: it cost a lot and she wants to protect it. She also identified another reason why it was important to keep her phone pristine. She explained that, when she was six years old, she had a strong yearning for that classic plastic toy, a Barbie doll – all her friends had one. She was given one as a present and really liked the ensemble of doll, shoes, clothes, accessories – just as it had come to her out of the box. Then one day she realized that one of the little plastic shoes was missing, and this fundamentally changed her relationship to the toy. She was no longer able to get the same pleasure from it. It no longer delivered on its much anticipated promise. It was spoilt. She feels the same way about her phone – if it were damaged, this would remove the possibility of a significant aspect of her 'user experience', her delight in the object, her joy in the potential for embodied extension into it. Because she knows that the inevitable signs of age – those tiny accretions of dirt and damage – are all too likely, she tries to forestall them to preserve the feelings of delight that she gets from her phone, even though doing so makes it impossible for her any longer to see or touch the surfaces she is protecting.

These seem relatively extreme cases, both of which probably fit Campbell's (1992) 'pristinian' type of consumer,[13] and they point to the degree to which the plastic surfaces of phone, case and screen protector that are in play in this new setting of the touch-screen device bring with them the qualities of the material with which we are familiar. The diffuse but powerful 'promise'[14] of pristine plastic surfaces is all too fugitive, and it may be necessary even to obscure it in order to preserve it, to have a valued object retain its value for us. It seems that the intimate relationship between touch-screen phone and body may raise concern for this loss of promise to a new pitch. Traces of this are evident even in the experience of individuals with less highly developed routines of preservation than the two people discussed above. Another interviewee explained that her rubbery phone case had got rather 'manky', and she had decided it was no longer acceptable and stopped using it, but she was conscious of the fragility of her iPhone, explaining that a friend had found a piece of glass in his mouth after his screen shattered. Touch-screen phones and their accoutrements therefore seem to share some of the repertoires of degradation that characterize other plastic objects, but these are activated in particular ways by the necessarily intimate way in which they are used. This may make plastics' palpable contribution to this new type of object rather different from its more familiar applications.

It would be possible to extend the principle of disgust (Fisher 2004) to cover this difference – certainly the propensity for the 'leakiness' of the body to elicit disgust is present in these new applications of plastics (Rozin 1999). However, there are aspects of these applications of plastic which have a significant bearing on individuals' attachment to their portable devices, influencing such issues as how and for how long they are kept – aspects that may not be covered adequately by a concept of disgust that emphasizes embodiment. Thus, the fact that phones and tablets *are* a portal to a network of communication – they function analogously as 'telescopes' as well as 'jewellery' – affords them a sense of agency that is not present in objects that are not 'wired'; after all, they talk back to us. They are entities that some of us take steps to protect despite the tendency of their surfaces to collect scratches, mould, chips, fragments of skin, hair and grease. These protective measures – entailing both bodily comportment and protective devices – point up the fact that these technologies are invested with a degree of 'life' akin to that which the interviewees may have found in their childhood toys. The implication is that what is being extended through such protective measures is not just a distributed body, but also an other, non-human agent the 'integrity' or 'untarnished identity' of which sustains our own.

Conclusion

As plastics break down, they betray the promises that they seem able to make when they are new and pristine. When new, plastics flatter our desire to remain in control of the material world, promising a sheer perfect surface for the world (to plasticize it). They appear to deliver this in the absolutely controlled engineered surfaces of consumer goods – forms facilitated by physical properties designed into the material, which can be specified to any task.

However, as this promise is betrayed, the magic is broken, the material can unsettle us and, in the process, we learn about its structure. The materials that make up touch-screen devices and the protective devices used to protect their surfaces bear traces of use in distinctive ways that highlight the physical interface of these portals to a networked environment. The margin between skin and plastic/metal/paint seems especially charged by being the point of contact between the material and the non-material aspects of communication technology. Because the examples discussed above concentrate on efforts to forestall this breakdown, they point towards moments when the positive feelings we may have about these elements of our 'extended selves' and our sense of self-and-other are destabilized, exposing the fragility of their material dimensions.

Notes

1 This seems a reasonable course, since many objects have only been widely available *in* plastic versions: ping-pong balls, hula hoops, beach balls, mobile phones, ball-point pens, CDs, telephones, condoms. To put 'plastic' in front of any of these creates a tautology because they are plastic objects by definition. While this chapter concentrates on encounters with palpable plastics in the present and moments when we might wish to discard individual objects, some classes of plastic objects have come and gone: cassette tapes, celluloid collars, cigarette holders and many others have suffered what amounts to 'species death'.

2 Thermoplastics absorb organic pollutions in the oceans – see Chapter 11 by Takada in this volume.

3 As the Society for the Plastics Industry puts it on its website: 'Plastics are responsible for countless facets of the modern life we enjoy today. From health and well-being, nutrition, shelter and transportation to safety and security, communication, sports, leisure activities and innovations of industry – plastics deliver bountiful benefits to you and your world. The men and women of the plastics industry make it all possible. In the United States, the plastics industry accounts for more than $374 billion dollars in annual shipments and directly employs nearly 1 million people.' www.plasticsindustry.org/aboutplastics/?navItemNumber=1008 (accessed 24 August 2012).

4 The presence in the environment and human bodies of plasticizing chemicals is linked to endocrine disruption, including birth defects in animals and humans. Concern centres on the Phthalates used to plasticize PVC and Bisphenol-A, which is a component of Polycarbonate (Freinkel 2011: 81ff; Vogel 2009).

5 Horst and Miller (2006: 61ff) researched the use of cell phones in Jamaica, and delineate the various ways in which they were worn on the body. Their work predates the introduction of touch-screen phones.

6 Another application of plastic films in this way, besides packaging, includes its use to protect dry-cleaning and car seats, and most recently to wrap luggage in air travel. In this use, sealing the suitcase in plastic ameliorates passengers' fear for the security of possessions that disappear into the baggage handling system. The visual effect of wrapping a suitcase in many layers of plastic film may also be relevant, as the film projects its characteristically soft but glassy sparkle to other passengers and, presumably, to baggage handling staff.

7 Seven exploratory interviews were conducted with touch-screen phone owners in 2011–12, with the participants selected for their strong level of engagement with phones in respect of the subject of this chapter. Four were phone shop employees

(two female, two male), three were female phone owners. These interviews were supplemented by some auto-ethnographic observation by the author (as a new touch-screen phone owner) and opportunistic informal enquiry with phone owners.

8 They are available even for phones that employ glass for the screen, which is very resistant to being scratched (though it is quite easy to shatter it) and they are manufactured to match both the front and the back of the phone, both of which may be made of glass.

9 The most complex of these has a double skin filled with the same polymer that is marketed as 'silly putty'. This is a non-Newtonian 'dilatant' fluid, because its viscosity increases with the force applied to it; here it is used as a shock absorber.

10 With the exception of cutlery that is disposable, which is often made of plastic.

11 This was an Internet survey of 28 males and 27 females. Of this self-selected sample, 31 were 44 years old or under and 24 were 45 or over. From this work, it is not possible to say whether the motivations for protecting phones are more likely to be economic or to do with the disquiet engendered by the degradation of the object's surfaces, but in combination with the interview this work shows that both motivations are in play.

12 This individual is a middle-aged female who works as an academic. Her commitment to preserving her possessions is likely to be stronger than most, but for this reason she was able to speak lucidly about the rationale for these care routines. She therefore constitutes an 'extreme' sample, useful because of the clarity with which she was able to speak about her habits in protecting her possessions.

13 Campbell's other types – the neophile and the technophile – would likely have a different relationship to the physical ageing of their possessions, possibly leading them to be less concerned about the signs of wear and degradation, since they represent motivations to acquire new goods on the basis that they are new, or more advanced, rather than because they are no longer pristine. Further work would be necessary to establish this.

14 Here, 'promise' is meant in the more restricted sense than Alison Clarke uses it in the title of her work on the history of Tupperware (Clarke 2001), to indicate the promise inherent in the qualities of plastic surfaces.

References

Banerjee, C. (2002) 'When All is not Fine with your Melamine', *The Tribune India*, online edition, 15 September, www.tribuneindia.com/2002/20020915/spectrum/sunday.htm (accessed 24 April 2012).

Barthes, R. (1987 [1957]) *Mythologies*, London: Paladin.

Belk, R. (2001) 'Possessions and the Extended Self', in D. Miller (ed.) *Consumption: Critical Concepts in the Social Sciences*, London: Routledge.

Benjamin, A. (2011) 'Endless Touching: Herder and Sculpture', *Aisthesis – pratiche, linguaggi e saperi dell'estetico* 3(1): 73–92.

Campbell, C. (1987) *The Romantic Ethic and the Spirit of Modern Consumerism*, Oxford: Blackwell.

——(1992) 'The Desire for the New: Its Nature and Social Location as Presented in Theories of Fashion and Modern Consumerism', in R. Silverstone and E. Hirsch (eds) *Consuming Technologies*, London: Routledge, 26–36.

Clarke, A.J. (2001) *Tupperware: The Promise of Plastic in 1950s America*, New York: Smithsonian Books.

Douglas, M. (1966) *Purity and Danger*, Oxford: Blackwell.

Falk, P. (1994) *The Consuming Body*, London: Sage.

Fisher, T. (2003) 'Plastics in Contemporary Consumption', unpublished PhD thesis, University of York.
——(2004) 'What we Touch Touches us: Materials, Affects and Affordances', *Design Issues* 20 (4): 20–31.
Fisher, T. and Shipton, J. (2009) *Designing for Re-Use: The Life of Consumer Packaging*, London: Earthscan.
Fredriksson, C. (1997) 'The Making of a Swedish Department Store', in C. Campbell and P. Falk (eds) *The Shopping Experience*, London: Sage, 111–35.
Freinkel, S. (2011) *Plastic: A Toxic Love Story*, New York: Houghton Mifflin.
Gibson, W. (1984) *Neuromancer*, London: Harper Collins.
Green, H. (2006) *Wood: Craft, Culture and History*, Harmondsworth: Penguin.
Horst, H. and Miller, D. (2006) *The Cell Phone: An Anthology of Communication*, London: Berg.
Ingold, T. (2010) 'Footprints Through the Weather-world: Walking, Breathing, Knowing', *Journal of the Royal Anthropological Institute* 16(S1): S121–39.
Keneghan, B. (2009) *Plastics: Looking at the Future, Learning from the Past*, London: Archetype Books.
McCracken, G. (1988) *Culture and Consumption: New Approaches to the Symbolic Character of Consumer Goods and Activities*, Bloomington, IN: Indiana University Press.
Meikle, J.L. (1995) *American Plastic: A Cultural History*, New Brunswick, NH: Rutgers University Press.
Michael, M. (2006) *Technoscience and Everyday Life*, Maidenhead: Open University Press.
Miller, D. (1987) *Material Culture and Mass Consumption*, Oxford: Basil Blackwell.
——(1994) 'Things Ain't What they Used to be', in S. Pearce (ed.) *Interpreting Objects and Collections*, London: Routledge.
——(2007) 'Stone Age or Plastic Age?' *Archaeological Dialogues* 14(1): 23–27.
Moravec, H. (1995) 'Bodies, Robots, Minds', www.frc.ri.cmu.edu/~hpm/project.archive/general.articles/1995/Kunstforum.html (accessed 1 July 2012).
Motamedi, N. (2007) 'The Aesthetics of Touch in Interaction Design', in *Proceedings of Designing Pleasurable Products and Interfaces*, 22–25 August, Helsinki, Finland.
Pepperell, R. (2005) 'Posthumans and Extended Experience', *Journal of Evolution and Technology* 14: 27–41.
Rankin, E. (1988) 'A Betrayal of Material: Problems of Conservation in the Constructivist Sculpture of Naum Gabo and Antoine Pevsner', *Leonardo* 21(3): 285–90.
Reckwitz, A. (2002) 'Towards a Theory of Social Practices: A Development in Culturalist Theorizing', *European Journal of Social Theory* 5(2): 243–63.
Rozin, P. (1999) 'Food is Fundamental, Fun, Frightening and Far Reaching', *Social Research* 66(1): 9–30.
Shove, E. and Pantzar, M. (2005) 'Consumers, Producers and Practices: Understanding the Invention and Reinvention of Nordic Walking', *Journal of Consumer Culture* 5(1): 43–64.
Vogel, S.A. (2009) 'The Politics of Plastics: The Making and Unmaking of Bisphenol-A "Safety"', *American Journal of Public Health* 99(S3): S559–66.
Webb, W. (2010) 'Being Mobile [Smart Phone Revolution]', *Engineering & Technology* 5(15): 64–65.

7 Reflections of an unrepentant plastiphobe

An essay on plasticity and the STS life

Jody A. Roberts

I'm looking at a picture of my daughter Helena when she was about six weeks old. It had finally become cold enough here in Philadelphia in late October for her to use some gifts she had been accumulating since before she was born. She's wearing a red hat covered with flowers, some cotton pants, socks that are barely hanging on to her feet, a couple of socks for her hands, and a blue hooded sweatshirt with the words 'FUTURE PLASTIPHOBE' emblazoned across the front. She also has a soft plastic tube extending out from her right nostril.

The sweatshirt referenced an inside joke. I had already earned a reputation among close friends as a 'plastiphobe' – a word we coined to describe my growing visceral reaction to anything plastic. When we received the gift, no one could have anticipated just how much plastic would be dangling from Helena in these early weeks of her life. The irony was not lost on my wife, Carrie, or me.

When Helena stopped breathing shortly after birth, our attempts to minimize contact with the synthetically produced and medicalized world – to live strictly in the world of the 'natural' – came to an abrupt halt. After she was intubated, the long plastic tube that ensured that air would continue to pass into her lungs symbolized our transition between these two different worlds. The proceeding weeks saw plastic tubing come and go. When intubation no longer seemed necessary, a nasal cannula carried fresh air into her nose, helping her to maintain an adequate supply of oxygen. Plastic tubes carried fluids flowing from plastic bags. In the early days, it was easy to stay distracted from the questions that occasionally crossed my mind. We were uncertain about so many things that I rarely had time to think about the environment of the neonatal intensive care unit (NICU), but, eventually, I began wondering: what other chemicals were entering Helena's fragile new body?

Reconfiguring our home to accommodate Helena's cerebral palsy has meant a continuation of many of the practices (and the presence of many of the materials that make those practices possible) that we first encountered in the NICU. Helena's plastic G-tube provides an alternative portal directly to her stomach. Every bit of nourishment that reaches her body passes through that plastic button and the plastic tube that connects it to syringes and enteral feeding bags. Now that she's no longer receiving breast milk (itself made possible by pumping through plastic flanges into plastic bottles and placed into plastic bags), each of her meals involves furious blending in a sturdy plastic blender.

My sense of place in this plasticized world was different before Helena was born. I was still living in a world synthetically fabricated from the miraculous molecules of the twentieth century, but things didn't feel quite so intimate. I had more control – or so I thought – to create boundaries between 'that' world and 'my' world. I didn't need to boil plastic bottles. No plastic tubing was required for eating. The days when Helena still received breast milk exclusively provided the perfect example of the shifting nature of the world I now inhabited: with loving and purposeful effort, Carrie and I worked to make sure that what she ate, and thus what entered her milk, remained as free of potential toxins as we could possibly manage, which meant (in part) avoiding plastic as often as possible; then we placed it into a plastic bag to be pumped through Helena's plastic tube into her stomach. This was the beginning of the end of my own techno-minimalist fantasies.

Up to now, I had been using my training in Science and Technology Studies (STS) to make better sense of the world around me; now I was turning these tools of inspection into tools of introspection, and in the process demonstrating the interconnections between the two. My theoretical approach is most heavily informed by three schools of thought: the social construction of technology (SCOT) programme; actor network theory (ANT) and related inquiries; and the so-called risk society.[1] My hope is that, in uniting these three frames of reference through the perspective of our own personal encounters, we might get closer to the discussions that lie beyond the facts and begin a conversation more deeply enmeshed in a politics of participation.[2] Drawing on this rich STS literature related to the construction of modern technologies, I hope this chapter helps to unite these various threads of scholarship in a way that highlights the roles we ourselves play in our everyday lives in the ongoing construction of society, and therefore our ability to take part in its reconstruction.

Becoming a plastiphobe

I am a plastiphobe – or at least I thought I was. At work, people would taunt me with plastic stirrers sitting in their freshly poured cups of hot coffee. They'd tell me how they just warmed up their lunch in a plastic container. They'd feign horror as they sipped from their plastic water bottles. After I delivered an informal lunch presentation about some of my research on emerging toxicological concerns related to chemicals, attention shifted away from the issue of plastics generally towards that of EDCs (endocrine disrupting chemicals) more specifically. They'd understood enough of the talk to find new ways to mock me. 'Uh-oh, Jody, are there any EDCs in this?' (Probably, but that's not why you're asking.) While they joke, the experience of being a plastiphobe is real. It becomes manifest in the things I see – the visible products I confront every day – as well as in what I can't see, but that I nonetheless know happens – the synthetic processes that make these products possible.

Staring at an individual object – or more likely dozens of those objects lined up together on a shelf – I can't help but sense the entire life-cycle of plastic, from extracted raw material (most likely petroleum or natural gas) to a destiny that will take it to landfill, incinerator or ocean. I stand staring at rows of items on store shelves, wondering which might be the least awful for me to purchase. I have a drawer in an office filing cabinet stuffed with plastic bags because I cannot bring myself to throw them away, but don't know what else to do with them because I can't be sure they are recyclable, or even if I try to recycle them whether or not they will indeed be recycled. When I look at a teabag steeping in a cup of hot water, I can't help but wonder what the bag is made of and whether or not it is breaking down in the presence of near-boiling water. When I plug in the humidifier in the bedroom to combat the dry heat of our radiators, I wonder what might be accompanying the steam as the water exits its polycarbonate container. I start to wonder where all of the finely ground pieces that once constituted the sole of my shoe end up after they have worn away. My concerns are not necessarily bound to plastics, but the category serves as a convenient place-holder for the questions and uncertainties about the chemicals of and in our modern world.

Human health and environmental health are typically dealt with separately. We have different regulatory agencies in the United States for addressing these problems. We have different academic departments, journals and degrees. It would seem as though humans live in another world, somehow altogether separate from the world that we actually inhabit. Synthetic chemicals – plastic or not – bridge these worlds. What is out there is now in us. Plastics might be poisoning us while they are also littering our landscapes, filling our landfills and polluting our oceans.

When academics get involved, these topics quickly become couched within a framework of uncertainty, risk and the politics implicit in both (Proctor and Schiebinger 2009). At stake is more than just a scholarly debate about epistemology. These are the elements of a lived reality (Callon *et al.* 2009). The debate for me is about more than plastic or pesticides; it involves navigating a way through a world made of plastic and made plastic by the artefacts of our technoscientific infrastructure. In our era of reflexive modernity, we must come to grips with a technoscientific world run amok (Beck 1992; Giddens 1991). Or, perhaps more appropriately, we must realize that we might be facing a runaway reaction. Unlike Beck or Giddens, though, I'm wrestling with my own participation in this system. I struggle to remake things before it is too late, even though I feel more enmeshed in the things to be remade than I ever have in the past.[3]

The emergence of plastics

The twentieth century bears the marks of a host of technoscientific triumphs over nature. Physicists, with the help of chemists, learned to smash atoms together, threatening both the built environment of human societies and the natural world. The emergence of modern genetics likewise provided new tools for technoscientists to construct new organisms with heightened precision and frightening amounts of unknown consequences. The century was also one in which the molecular make-up of the world was irrevocably altered in a multitude of ways. The growth of the organic chemicals industry, developed in the nineteenth century to create new synthetic dyes and fertilizers, changed gear; as nations went to war, so did chemists. Pesticides and poisons (one and the same, really) became mainstays of the chemical research lab (Russell 2001). While chemists in labs were creating new molecular substances, chemical engineers were developing more efficient ways of mass producing them (Ndiaye 2007). With the war machine consuming nature's bounty, and with U-boats blocking the global trade so vital to wartime production, chemists expanded their research into the production of synthetic molecules for the replacement of nature's goods. New plastics, which had familial relations to the early compounds that went by the names of celluloid and Bakelite, were of a different sort. They came in infinite varieties, could be put to use in a multitude of situations, and most importantly could be created cheaply in tremendous quantities. The flow of plastics did not abate after the war; it simply entered new markets. People produced more polymers, and the plastics kept piling up. That's the world we live in now: a plastic paradise.[4]

Plastics, or any of the molecular markers of the modern chemical industry, don't stay put. They are unruly technologies. Recent advances in human biomonitoring studies conducted by the US Centers for Disease Control and Prevention (CDC) have sought to quantify, in some statistically probable kind of way, the level of exposure to synthetic chemicals faced by everyday Americans. As new techniques are developed, the list increases of chemicals sought out, and found, in our blood, urine, fat, breast milk, and other organic tissues and fluids (CDC 2009).[5] As more data become available, the CDC has begun supplementing previously published reports with new results, including one for Bisphenol-A (BPA). Over the last few years, BPA has become the favourite battleground for industry, health and regulatory wrangling because of its pervasiveness in the marketplace and its specific uses in products designed for children (such as baby bottles). The results of these new tests, conducted between 2003 and 2004, found BPA in 92.6 per cent of all people tested (Calafat *et al.* 2008). Given the uses of the chemical, this finding was perhaps not terribly surprising.

The debates about BPA, however, are just a microcosm of the larger landscape that includes the molecular reconstruction of our environments, the implications this may have for our bodies and those of the organisms around us and in us, and how best to govern these processes. Hundreds of industrial chemicals – from plasticizers and pesticides to flame retardants and solvents – are now reliably located in the bodies of every resident of the United States. However, this fact has been difficult to translate into other contexts, particularly regulatory ones.[6] The CDC (2005) keeps quiet about possible links between its biomonitoring data and possible health consequences because, by and large, no one has ever looked for any connections. Industry advocates emphasize the same lack of correlative evidence.[7] However, while the CDC has been careful not to overstate the case, and industry advocates have systemically understated it, environmental and health-focused organizations have used the data (from the CDC as well as self-generated) to raise a host of concerns.[8]

Another way of putting it is to say that, while the CDC and industry activists may argue that proof of presence does not equal proof of harm, proof of non-harm (such that it is possible) has also been elusive.[9] At this point, it may even be difficult to understand what that harm will look like. Will it come in the form of cancer, obesity or infertility? When will these effects become manifest: late in life, in my children, in my children's children?[10] While we may not know, the fact that we are asking these questions is an indication in itself that some knowledge has emerged and that the politics of plastic are changing. The molecular transformation of the earth's ecosystems that is currently taking place should at the very least be a matter of concern for all of us, especially when it comes to protecting the most vulnerable populations – like my daughter.

I don't know much about developmental embryology, but I do know that it is complicated, a carefully integrated process of cell growth, expansion and specialization. At the root of much of this process is the constant chemical wash the foetus experiences: hormones flowing in from the mother and soon from its own developing endocrine system. BPA, phthalates, organo-phosphate and organo-chlorine pesticides, polybrominated diphenyl ether (PBDE) and other flame retardants, polychlorinated biphenyls (PCBs), dioxins and other chemicals encountered on a daily basis have all been shown to disrupt this delicate and absolutely crucial process to varying degrees. However, understanding what this means for my – our – lives is much less clear. How should I change my daily routine? Would it matter? What would it take to extract myself from the fabric of our society? Perhaps more simply, how are we to live? It is this last articulation that is most important, since it highlights the complexity of the dilemma. This is not only or merely a scientific issue, one of uncertainty about data related to health. These questions uncover with deep intimacy the relations between the ethical structure – or perhaps grammar – of the technological system that produces these artefacts. My choice to participate or not (or to what extent) is precisely the sort of ethical wrangling that is at work in, say, Peter Singer's (1995) musings on the same question.

A plastic pregnancy

If friends and colleagues had found me a little unstable for my reactions to all things plastic already, the experience of Carrie's pregnancy only intensified the situation. Items and actions that previously had been acceptable were suddenly excluded. Polycarbonate bottle? Off limits. It was replaced with an unlined stainless steel container. Strawberries? Out. They're nothing but congealed pesticides. Navigating our way through a plastic-laden world had always been difficult, but now – with even more at stake – the task became all-consuming.

The peak of frustration came roughly seven months into the pregnancy. Carrie, exhausted from sleepless nights and back pain, determined one Saturday morning finally to purchase a new mattress. With literally one foot out of the door, I froze: what about the flame retardants – the PBDEs (one of a class of widely used flame retardants)? Study after study found them to be potent endocrine disruptors. Here in the United States, all furniture (especially mattresses) must meet strict fire-code regulations that result in the dousing of the lining and stuffing within with flame retardants such as PBDE. Why? It would seem that there are fears that I might light the house on fire if I fall asleep smoking in bed, but my fear had nothing to do with smoking in bed or dying in an inferno as a result of something so far removed from my everyday life. Instead, I pictured my head, Carrie's head and the yet-to-be-seen head of our baby lying comfortably together on this new bed, rhythmically inhaling the flame retardants off-gassing into our bedroom. I couldn't go. I couldn't even bring myself to walk out the door. Carrie, now nearly in tears from disappointment and frustration, left it to me to find a way through this tangled mess of knowing and unknowing. Knowledge (or uncertainty, or un-knowing) was not enough; an action needed to be taken to resolve a real, not hypothetical or probabilistic, issue. I had to decide.

I spent the day searching for alternatives, trying to balance what I thought would keep us safe with what I could realistically afford to spend, all enmeshed in the fabric of a cultural and legal network of codes that sought to protect me. Yet I was in pursuit of the same thing: safety, but with constant aggravation at the ways the approaches constrained one another. It was precisely this network of laws, regulations, codes and material substances that prevented me from doing what I thought to be safest.

I was, admittedly, attempting to shop my way to safety (Szasz 2009). I understood that, but what alternative did I have? I could write to my congressional representatives asking for reforms to be made to the federal statutes governing my exposure to these chemicals. I could write an op-ed about the persistence of outdated regulations and the unforeseen public health problems they had created. I could attempt to research and write the history of fire codes in the United States. I could trace the parallel developments in organic chemistry that allowed those codes to be met. I could look for ways to link the ways in which the persistence of those chemicals and those practices had now potentially created a problem far more pervasive than the original problem. Or I could try my best to buy my way out of the problem. So I did. I ordered a wool-filled mattress topper (wool is by nature flame retardant, and so meets US fire codes) that would temporarily relieve us of the uncomfortable nights without having to purchase something doused in synthetic flame retardants, which would simply have substituted Carrie's sleeplessness for my own.[11]

After navigating 40-plus weeks of pregnancy, we found ourselves at the threshold of two converging worlds. What had been inside was now ready to be outside. What had been unseen for so long, but sensed nonetheless, would soon become visible. The skills we carefully honed during the pregnancy would have to persist across this divide. We would need to protect the milk from the same substances that we tried to keep from passing through the placenta. We would take the same care working in the outside world as we'd taken when dealing with the inside one. We'd been so careful, so vigilant, so exhaustive that we were sure that we'd done everything we could to protect this fragile developing life from the toxic world it was about to enter. When Helena finally emerged, crying and groping for food and warmth, we thought we had succeeded – until she stopped breathing.

The NICU is an otherworldly place (Layne 1996). Given the seriousness of many of the situations that characterize that space, it is remarkably quiet and dark (light becoming a sign that something might be wrong and that greater observation is required). Two things, however, were omnipresent in this new environment: noise (the various beeps associated with a sundry of monitoring equipment) and plastic.

It took about 10 days for me finally to acknowledge that I was living in a plastic nightmare. The majority of the babies there were in incubators, like those seen in an American Chemistry Council 'Essential$_2$' ad campaign. Luckily, Helena was not, but she was covered in soft, flexible medical tubing – the kind infamously full of phthalates. Nearly everything that came into contact with her passed through this tubing. Plastic is the preferred material in that environment. Plastic held her growing supply of breast milk, collected with the hope that she'd soon be able to ingest it; plastic bladder bags held the variety of liquids that were slowly pumped into her; plastic comprised the walls of the bed upon which she lay motionless in those first few weeks.

Manifestations of my plastiphobia took two forms. Most immediately, I found myself wondering what was in the particular plastics that were attached to or literally in Helena. *Which* phthalates did they use to make that tubing so pliable? How likely were they to leach when liquids were passed through them? As Helena moved from IV to nasal tube for eating, I tried to imagine the soft tubing dangling down her throat, precisely placed just inside the opening of her stomach. How resistant would this tubing be to the acidity of the stomach? Would her body attempt to digest the tube – this foreign but essential device that helped to keep her alive?

I tried to talk to some of the nurses about my concerns, but only casually. From my previous experiences, I knew that most people would not welcome this line of questioning. Emotionally, I was not in a place to take the gentle ribbing to which I had grown accustomed at work or at home. I succeeded with them about as much as I did with my co-workers: they knew just enough to take part in some slight mocking of my concerns, but without any real concerns of their own.

However, it wasn't just the plastic (potentially) accumulating in Helena's body that bothered me; it was all of the plastic flowing into the world – accumulating in large heaps in the trash cans that were emptied several times a day. These mounds of plastic that each day and night filled the trash cans in our one tiny corner of suburban hospital-land left me feeling strangely deflated and defeated. It was as if that assault on the earth was just another injury being perpetrated on Helena. Plastic syringes, bladder bags, pre-measured and mixed formula, tubing and all the other plastic miscellanea all headed to another location where the accumulated matter would likely be incinerated, which itself would lead to the production and accumulation of new compounds in that environment before molecules of dioxin and other persistent organic pollutants would ride the currents of water and wind to the coldest locations on earth to be deposited, consumed and deposited again in the fat of a seal, polar bear or human being.

Becoming plastic

In his essay 'Plastic', Roland Barthes (1972) gets at the fundamental problem of trying to deal with these mysterious substances. As a substitute for all things, plastic is everything. As raw matter that can become anything, is it really anything itself? Plastic is troubling precisely because it lays old categories to waste. In the half-century since 1957, when Barthes originally wrote his essay, we've witnessed plastic's emergence in nearly every aspect of our everyday lives. From transportation, to food, to clothing, to shelter, we live in a thoroughly plasticized world – even if this real world is not quite the one envisioned by the 'plastic pioneers' of the past (see Morris 1986; Meikle 1995). The spread of plastic has been more subtle, and it is perhaps for that reason that experts of all stripes missed it slipping into unintended places, travelling near and far such that nearly every cup of water from the ocean is likely to contain some plastic in some form of degradation and nearly every human subject found anywhere on the globe will likely bear the marks of a plastic modernity. What, though, are these plastics doing to us?

Twenty-seven days after Helena was born, we were discharged from the NICU. The car was packed with little more than Carrie, Helena and me, but the reconstruction of our apartment into a mini-NICU had already taken place. Waiting for us at home were apnoea and pulse/oxygen monitors, two tanks of oxygen, an air pump, an oxygen concentrator and a month's supply of plastic tubes and syringes for feedings. Our one-bedroom apartment was suddenly filled with boxes, bottles and beeping. This was worse than the NICU, where at least the stimuli could be isolated and more easily ignored. Now, with every beep, we went running, tripping over cords and tubes, trying not to spill the precious breast milk.

When we had previously inhabited our home, nearly everything was designed to minimize our, and even more so Carrie's, contact with plastic – especially when it came to food. Part of this was for the pregnancy – for Carrie's health and the health of the developing foetus – but it was also, in part, a preparation for life after the birth. With Carrie nursing (at least that was the plan), we wanted to eliminate as much as possible our contact with potential contaminants. However, for the past four weeks we'd been eating almost nothing that *didn't* come from a plastic container: #5 containers filled with goodness from my father-in-law's café in Annapolis, cookies wrapped in plastic wrap, prepared salads in plastic clamshells, paper cups, plastic juice bottles – the list goes on. The very network I had been working to deconstruct, or perhaps more accurately reposition us outside, was now the network within which I subsisted. It sustained us just as it had sustained Helena during those crucial hours and days, and would continue to do so in the time yet to be defined.

While I had hoped to avoid the excesses of a plastic modernity when we left the hospital behind, I instead found myself recreating many of the habits and rituals learned in those early days in my home, albeit now moulded and modified (like a new plastic) to reflect my own rituals adopted from our lives before and made as fitting as possible for our new experiences. I find myself flushing the brand-new 'sterile' bladder bag and tube with water before filling it, the aroma of fresh plastic (typically a result of phthalate monomers off-gassing) filling my nose. With that smell, I recall a story from a recent environmental endocrine disruptor conference I had attended. I was sharing stories with a couple – recent parents themselves – and the mother had told me how she had her partner flush the tubes in the hospital before she was hooked up to an IV prior to her delivery. In that moment, with that smell, I had suddenly assumed that I should be doing the same thing – that I should *always* have been doing the same thing – but what difference would it really make?

For better or for worse, our lives are dominated and made possible in a tangible and real way by the very plastic that I've tried so hard to avoid. It's the bottle that held the breast milk; the tubes and flanges that made pumping possible; the bladder bag that held the milk to be delivered; the tube that connects the bag to Helena; and of course the G-tube (and previously the NG-tube) that provides access to Helena's stomach.

I'm not only concerned about the plastic that sustains our lives, but also about the quantities that daily I'm asked to dispose of in no particularly good way. Opening the trash can, I can't help but think of that strange phenomenon variously known as the Pacific Trash Vortex, Great Pacific Garbage Patch, the Eastern Garbage Patch or simply the Plastic Soup.[12] At the spot in the Pacific Ocean where the currents merge and begin to swirl in on themselves, Charles Moore (founder of the Algalita Marine Research Foundation) has found an island of trash that currently stands at roughly twice the size of Texas (though every time I research the phenomenon, the size seems to have grown). The 'island' is mostly a soupy mixture of all things plastic – the indestructible element of our lives.

Plastics are not static molecules. They are not easily tamed and do not stay put. They are, as I noted earlier, unruly. Molecules lost during production or simply broken away become mobile. Some plastics travel thousands of miles through earth, wind and water, slowly breaking down, polymer dissolving ever so slowly towards monomer and spreading through the food chain; some have simply ridden the currents to this trash dump at sea. The trash that collects here serves as a dumping ground for our culture, and as a feeding pit for the locals. Seabirds, turtles and fish fill their bellies with the waste. Those not overwhelmed and taken over by the sheer quantity of the plastic they consume will instead become part of the plasticized food chain. Staring at the images of this dump, I realize that I'm sitting in the centre of it, resting not so comfortably amid the shopping bags, diapers, toys, and all of the other plastic miscellanea that are our culture's gift to the world.

The plastics that populate my everyday life and that fill me with such anxiety also help to make Helena's life possible, but we resist the simple dichotomies imposed on us. The plastics are not simply life saving or a threat: they are both. At the same time, my lack of knowledge is not a sign of deficit, but a sign of the shifting relationships of my life and world. More importantly, we can shape the direction the plastics take. We can redesign them to be more benign. We can protect vulnerable populations from exposure. We can decide how and when and why we use these materials. Their future is as much unwritten as our own. Together, we are becoming plastic.

Acknowledgements

A previous version of this essay was published in *Science as Culture* as part of a special issue, 'Embodying STS: Identity, Narrative, and the Interdisciplinary Body' (2010, 19(1): 101–20). The author thanks Taylor & Francis for permission to print this adapted version. The author also wishes to express his thanks to the editors of the present volume for allowing him the opportunity to expand on and refine some of those ideas for this collection.

Notes

1 For foundational works in the social construction of technology, see Bijker, Hughes and Pinch (1987). Exemplary (and relevant) works in the tradition of actor network theory include Latour (1987, 2004), Mol (2002) and Callon (1987), which is also an interesting example of when SCOT almost met ANT. The foundational work in STS for thinking about the risk society is Beck (1992).

2 See Cohen and Galusky (2010) for a discussion of STS as a participatory activity.

3 For insight on Beck's take on life in this reflexive modernity, listen to his interview on 'Ideas: How to Think About Science', www.cbc.ca/ideas/episodes/2009/01/02/how-to-think-about-science-part-1—24-listen/#episode5 (accessed 13 July 2012).

4 For this brief overview of the history of plastics, I've drawn on Meikle (1995).

5 The 2009 report is the most recent *complete* report, but the CDC has continued to release updates (including in 2012), demonstrating expanded surveillance of specific chemicals and new human sources, all of which can be found on the CDC website: www.cdc.gov/exposurereport (accessed 13 July 2012).

6 For an overview of some of the difficulties of implementing these data, see NRC (2006) and Roberts (2008a). For the difficulties of fitting this work into legal and regulatory frameworks, see Cranor (2008) and Roberts (2008b).

7 See the American Chemistry Council's page on biomonitoring: www.americanchemistry.com/Policy/Chemical-Safety/Biomonitoring (accessed 13 July 2012).

8 See, for example, the Environmental Working Group's 'Human Toxome Project': www.ewg.org/sites/humantoxome (accessed 13 July 2012). See also Lang *et al.* (2008).

9 There are two ways of dealing with this situation. In the first instance, we might broadly classify this as an example of 'agnotology' – a case of things we simply don't know (Proctor and Schiebinger 2009). However, why don't we know? Is this unknowable (and therefore ignorance is unavoidable), or is it a case of undone science (with all of the politics that implies)? See Frickel *et al.* (2010) for a discussion of the latter proposition.

10 These are just some of the questions with which researchers working in these areas are grappling, and also just a small sample of a growing list of possible exposure endpoints. The case around BPA serves as the quintessential case.

11 We repeated much of this same ritual a year later when we moved, in search of a larger bed to accommodate two adults, a toddler and two cats (one of which insisted on keeping his place at my feet despite the altered sleeping conditions). The problems persist nearly as much as the chemicals themselves.

12 See, for example, the Greenpeace website: www.greenpeace.org/international/campaigns/oceans/pollution/trash-vortex (accessed 13 July 2012); and Marks and Howden (2008).

References

Barthes, R. (1972 [1957]) 'Plastic', in R. Barthes, *Mythologies*, trans. A. Lavers, New York: Hill and Wang, 97–99.

Beck, U. (1992) *Risk Society: Towards a New Modernity*, London: Sage.

Bijker, W.E., Hughes, T.P. and Pinch, T.J. (eds) (1987) *The Social Construction of Technological Systems*, Cambridge, MA: MIT Press.

Calafat, A.M. *et al.* (2008) 'Exposure of the U.S. Population to Bisphenol A and 4-tertiary-octylphenol: 2003–4', *Environmental Health Perspectives* 116(1): 39–44.

Callon, M. (1987) 'Society in the Making: The Study of Technology as a Tool for Sociological Analysis', in W.E. Bijker, T.P. Hughes and T.J Pinch (eds) *The Social Construction of Technology*, Cambridge, MA: MIT Press.

Callon, M., Lascoumes, P. and Barthe, Y. (2009) *Acting in an Uncertain World: An Essay on Technical Democracy*, Cambridge, MA: MIT Press, 77–97.

Casper, M.J. (ed.) (2003) *Synthetic Planet: Chemical Politics and the Hazards of Modern Life*, New York: Routledge.

CDC (Centers for Disease Control and Prevention) (2009) *Fourth National Report on Human Exposure to Environmental Chemicals*, Atlanta, GA: CDC.

Cohen, B.R. and Galusky, W. (2010) 'Guest Editorial', *Science as Culture* 19(1): 1–14.

Cranor, C.F. (2008) 'Do You Want to Bet Your Children's Health on Post-market Harm Principles? An Argument for a Trespass or Permission Model for Regulating Toxicants', *Villanova Environmental Law Journal* 19(2): 215–314.

Frickel, S. *et al.* (2010) 'Undone Science: Social Movement Challenges to Dominant Scientific Practice', *Science, Technology, and Human Values* 35(4): 444–73.

Giddens, A. (1991) *Modernity and Self-Identity: Self and Society in the Late Modern Age*, Stanford, CA: Stanford University Press.

Haraway, D. (2008) *When Species Meet*, Minneapolis, MN: University of Minnesota Press.

Jonas, H. (1984) *The Imperative of Responsibility: In Search of an Ethics for the Technological Age*, Chicago: University of Chicago Press.

Lang, I.A. *et al.* (2008) 'Association of Urinary Bisphenol A Concentration with Medical Disorders and Laboratory Abnormalities in Adults', *Journal of the American Medical Association* 300(11): 1303–10.

Latour, B. (1987) *Science in Action*, Cambridge, MA: Harvard University Press.

——(2004) *Politics of Nature: How to Bring the Sciences into Democracy*, Cambridge, MA: Harvard University Press.

Layne, L. (1996) '"How's the Baby Doing?" Struggling with Narratives of Progress in a Neonatal Intensive Care Unit', *Medical Anthropology Quarterly*, special issue 'Biomedical Technologies: Reconfiguring Nature and Culture', ed. B. Koenig, 10(4): 624–56.

Marks, K. and Howden, D. (2008) 'Vast and Growing Fast, a Garbage Tip that Stretches from Hawaii to Japan', *The Independent* (London), 5 February, News: 2.

Meikle, J. (1995) *American Plastic: A Cultural History*, New Brunswick, NJ: Rutgers University Press.

Mol, A. (2002) *The Body Multiple: Ontology in Medical Practice*, Durham, NC: Duke University Press.

Morris, P.J.T. (1986) *Polymer Pioneers: A Popular History of the Science and Technology of Large Molecules*, Philadelphia, PA: Beckman Center for the History of Chemistry.

Ndiaye, P. (2007) *Nylon and Bombs: DuPont and the March of Modern America*, Baltimore, MD: Johns Hopkins University Press.

NRC (National Research Council) (2006) *Human Biomonitoring for Environmental Chemicals*, Washington, DC: National Academies Press.

Pickering, A. (1995) *The Mangle of Practice: Time, Agency, and Science*, Chicago: University of Chicago Press.

Proctor, R. and Schiebinger, L. (eds) (2009) *Agnotology: The Making and Unmaking of Ignorance*, Stanford, CA: Stanford University Press.

Roberts, J.A. (2008a) 'New Chemical Bodies: A Conversation on Human Biomonitoring and Endocrine-disrupting Chemicals', *Studies in Sustainability*, Philadelphia, PA:

Chemical Heritage Foundation, www.chemheritage.org/pubs/New-Chemical-Bodies.pdf (accessed 13 July 2009).

——(2008b) 'Collision Course? Science, Law, and Regulation in the Emerging Science of Low Dose Toxicity', *Villanova Environmental Law Journal* 20(1): 1–21.

Russell, E. (2001) *War and Nature: Fighting Humans and Insects with Chemicals from World War I to Silent Spring*, Cambridge: Cambridge University Press.

Singer, P. (1995) *How Are We to Live? Ethics in an Age of Self-interest*, New York: Prometheus.

Szasz, A. (2009) *Shopping Our Way to Safety: How we Changed from Protecting the Environment to Protecting Ourselves*, Minneapolis, MN: University of Minnesota Press.

8 Plasticizers

A twenty-first-century miasma

Max Liboiron

Some Greenland natives have such high quantities of industrial chemicals in their bodies – including those used in plastics – that they can be classified as toxic waste when they die (Cone 2005: 32). Scientists have found that every person tested in the United States and Canada, and many other countries, carries plastic chemicals in his or her body (CDCP 2009; Bushnik *et al.* 2010). These plastic chemicals have complex and largely uncharted effects. Plastic pollution is unique not only because of its ubiquity and persistence, but also because most efforts to solve or mitigate its effects are failing. Typical proposed solutions to bodily chemical burdens include banning bisphenol-A (BPA) in baby bottles, avoiding plastic food containers and increasing recycling. Yet, if all plastics were recycled, all plastic food containers were eliminated and all demonstrably harmful chemicals were banned, bodies would still not be free of plastic pollution.

In this chapter, I argue that miasmas can be useful for thinking through plastic pollution because the miasma theory shares several characteristics with the behaviours of plastic chemicals. Miasmas may appear to be an unscientific and folksy concept, but they were the first and longest-running scientific theories of disease that did not attribute illness to spiritual causes. They were also logical models used to explain dispersed, unspecific influences within the wider environment that caused bodily harm and illness. I suggest that these characteristics have acquired new relevance for examining plastics and their behaviours.

Before the late nineteenth century, disease was perceived to be caused by 'bad air', or miasmas. By the turn of the twentieth century, the miasmic model of harm had been replaced by germ theory. Within 40 years, a model of pollution developed that privileged linear causal links between a discrete pollutant and its pollution, and the quantification of harm.

Miasmas exemplify what I call the influence model of harm, in contrast to the particle model of harm used today that describes the actions of discrete pollutants. Models, as Mary Morgan and Margaret Morrison (1999) explain in *Models as Mediators*, 'provide us with a tool for investigation, giving the user the potential to learn about the world or about theories or both because of their characteristics of autonomy and representational power, and their ability to effect a relation between scientific theories and the world' (Morgan and Morrison 1999: 35). Models hold together disparate facts to describe how different parts of the world relate to each other. Their representational power proposes not only how the world works, but also how particular solutions follow from specific representations; they both explain and enact the world around them. Miasma and the influence model 'effect a relation' between environments, bodies and ill-health that describes plastic body burdens better than current particle models of harm, which consistently fail to describe or control plastic pollution.

This chapter first outlines the miasma theory of disease, the influence model of harm and our current model of pollution, focusing on each model's internal architecture and physical form; its agency, or mechanism of harm; and its geography, or the spatial relations within bodies and the spaces outside of them. It then compares the two models with the phenomenon of plastic body burdens. Each model generates modes and points of intervention, and I argue that the miasma theory provides better representations, and thereby more appropriate interventions for mitigating the effects of bodily plastic pollution, than current dominant models and solutions.

Miasmic logic

Between the sixth and early twentieth centuries, the miasma theory of disease posited that diseases, and particularly epidemic illness, were caused by 'an ill-defined but universally recognized corruption and infection of the air' (Cipolla 1992: 4). This infection of the air originated from such diverse sources as extreme weather, decaying organic matter, corpses, marshes, cess-pools, depressions in the earth, volcanic eruptions, human exhalations and the conjunction of the stars. External causes of illness, such as miasmas, were called the exciting or immediate cause of disease – they were not the disease itself, but could cause an imbalance of humours within a body, which in turn could result in illness. Both plastic chemicals and miasmas operate through what I call the influence model because the architecture of miasmas – particularly their inextricability from the surrounding environment, their mechanisms for 'causing' illness and even the definition of illness they helped to explain – pivots on a certain mode of effect, an 'action or fact of flowing in' of something amorphous but forceful, through a 'secret power or principle'. Influence is the 'exertion of action of which the operation is unseen or insensible' (*Oxford English Dictionary* 2012), and as such is appropriate to miasma and its form and agency.

Miasmas were inextricable from the landscape, urban architecture and the human population. Their mechanism of harm was not direct, but additive and somewhat mysterious: weather, personal histories, architecture, diet, the alignment of stars, the location of cesspools, plumbing and employment conditions all had to be counted by physicians trying to cure the sick, and by sanitarians aiming to reduce the presence of miasmas in their locales. As such, miasma's epistemological structure was one of accumulation, where everything was taken into account and nothing could be dismissed or overlooked (Latour 1988). Moreover, miasmas were not lone agents of illness. They coexisted with other theories of disease, such as contagion or infection. Some present-day authors have argued that 'early nineteenth-century physicians who explained disease by too many causes (miasma, tainted water, contaminated soil, poor living, sleeping while intoxicated, etc.) actually had no causal theory at all' (Kern 2004: 74). This view is suited to the ways in which we think of causality today, rather than how causality operated via influence in earlier centuries. It was not that the hygienists and miasmaists failed to locate and correlate causes for specific diseases; rather, it was the nature of miasma to be multidirectional, variegated and indiscriminating, and to work in concert with other mechanisms of illness. Experts thus had to account for an ill-defined, radically irreducible phenomenon.

The first 100 pages of *The Uses and Abuses of Air*, a treatise written by John H. Griscom (1854), an American miasmaist and leader of the sanitation movement in the mid-nineteenth century, compares breathing to food, clothing and other basic human needs. The treatise goes on to describe how 'pure' and 'bad' air interacts with digestion, blood circulation, the liver, heart, lungs, skin, ears and eyes. Griscom explains ties between air, vice and occupational setting. Whether he is discussing 'the amount of water thrown off from the lungs' or how 'vitiated air produces quarrelsomeness and encourages intemperance and other vices', bad air and harm are never studied in isolation. They are not separate from the body, the wider atmosphere, buildings or moral behaviours, and they potentially affect all bodily systems. In short, illness and its causes were systemic rather than discrete, holistic rather than piecemeal.

The logic of influence and layers of causality explained how individuals could resist certain diseases, why overworked and exhausted people became ill more often and with more dire results, and why one member of a household could resist disease or recover from disease while the rest fell ill or died. In his treatise on miasmic air, Griscom (1854) tells the story of three men sharing a sofa in a crowded room. It is winter, the windows are shut and the room becomes impure with exhalations:

> The first-mentioned [gentleman] complained of the *great heat* of the room, and wished the fire might be extinguished; the second said he felt *very cold*, even in that high temperature, and actually rose and *put on his greatcoat*; while the third, as if mysteriously influenced by the two extremes by which he was flanked, complained of alternating feverishness and chilliness, and expressed himself unable to account for the singular sensations. Thus we are given an exhibition of the manner in which different temperaments are affected by one cause [the density of carbon dioxide in crowded rooms].
>
> (Griscom 1854: 57–58, emphasis in original)

There is no linear or guaranteed relationship between the presence of a potential harm and the harm it produces. Instead, countless variables, presupposing causes and unique contexts both within and outside of bodies join to produce a specific effect. Just because a condition of harm is present – in Griscom's case, elevated levels of carbon dioxide and reduced levels of oxygen – illness does not necessarily follow, nor is the manifestation or type of harm the same for all people exposed.

This is not to say models of influence produce random phenomena. Sanitarians such as Griscom noticed that certain predisposing conditions, such as poverty and habitation in tenements, consistently led to illness and epidemics. Griscom revolutionized public health because he advocated that it was the surrounding conditions of poverty, rather than people living in poverty, that led to epidemics and chronic illness – that is, poverty was an immediate cause, rather than the more popular theory that predisposing causes such as immorality made tenement dwellers both poor and ill. At the time, this was a radical reorganization of causality. His brand of environmental health was an early form of environmental justice, though such a term would not exist for more than a century.

The autonomous pollutant

The germ inspired a radical new model of harm starting in the 1880s that would develop over the course of the twentieth century. With the germ theory, illness and health changed from a confluence of environment, body and morality to a single, discrete pathogenic agent. This affected both the form and mechanisms of disease. In 1882, Robert Koch, one of the founders of germ theory, wrote:

> Only a few decades ago the real nature of tuberculosis was unknown to us; it was regarded as a consequence ... of social misery, and as this supposed cause could not be got rid of by simple means people relied on the probably gradual improvement of social conditions and did nothing. All this is altered now. We know ... the real cause of the disease is a parasite – that is, a visible and palpable enemy with which we can pursue and annihilate, just as we can pursue and annihilate other parasitic enemies of mankind.
>
> (Koch 1961: 461)

Koch's experiments are characterized by an ethos of isolation. Pathogens were separated from the body and the environment on Petri dishes in laboratories. These 'visible and palpable' enemies were things-in-themselves, separate from the symptoms they caused in bodies. The isolation of causality to one variable is a radical departure from the systemic influence model of disease based on disruptions and balance (Nash 2006). The autonomy of germs allowed a simultaneous segregation of bodies, environments and disease, reducing or eliminating the former system of co-dependence and influence.

Doctors could now focus their attention on germs rather than an endless plethora of influences. In 1916, Hibbert Winslow Hill, director of the Institute of Public Health in Minnesota, wrote: 'The old public health was concerned with the environment; the new is concerned with the individual. The old sought the sources of infectious disease in the surroundings of man; the new finds them in man himself' (Hill 1916: 8). The germ changed concepts of disease and health. Health came to denote the absence of pathogens rather than an imbalance of forces. The environment, together with other 'action[s] of which the operation is unseen or insensible', was no longer central to the disease equation.

The term 'pollutant' was not documented until the end of the nineteenth century, seven centuries after the word 'pollution' appeared (*Oxford English Dictionary* 2012). There is no entry for pollutant in Griscom's *The Uses and Abuses of Air*, or any text prior to 1892 (n.a. 1892). Its emergence coincides with the acceptance of the germ theory of disease and the model of isolated, harmful agents. Pollutants and germs were not and are not synonymous, but pollutants and germs did evolve concurrently and share an overall model of discreteness and segregated architecture, as well as linear causality. I will call this the particle model of harm.

Within the particle model, the relationship between a disease – now a particular disease rather than disease writ large – and its cause are linear. One of the primary tasks that Koch and the new generation of microbiologists pursued was discovering a microorganism's law-like rules of behaviour and pathways of movement. Germs did not reappear and disappear like miasmas. Rather, they moved dependably from point A to point B to point C. Instead of correlating countless everyday sources of decay, from exhalations to puddles, weather, life events and astronomy, these new microorganisms allowed scientists to 'follow the thread' of locomotion that emerged from the essential character of each individual disease (Latour 1988: 46). Now causes of illness had 'essential' characters, not amorphous natures.

While the germ and miasma theories coexisted for nearly 20 years at the turn of the twentieth century, there was debate over the advantages and disadvantages of each model. The separation of the larger environment, occupational and living arrangements, poverty and vice from disease were the main sources of conflict between miasmaists and germ theorists. The new form, agency and geography of germs led to a new type of disease intervention. Rather than making changes in the environment, inoculations and other disease-prevention measures focused exclusively on the microbe. According to miasmaists, the germ theory did not adequately explain why the poor and those employed in the hardest conditions of living and labour were disproportionately affected by disease (Elliot 1870: 488–89). Nevertheless, the germ theory has become the dominant model of harm used in disease prevention and pollution control in the twenty-first century. However, while this particle model has served us well for the last century, and has met with great success, it fails to account for some types of disease and chemical harm, such as those associated with plastic pollution.

Plastic pollution

In 1941, as the mass production of plastics was just beginning, V.E. Yarsley, a chemist, and E.G. Couzens, a research manager for B.X. Plastics Ltd, extolled the virtues of a new plastic world:

> 'Plastic Man,' will come into a world of colour and bring shining surfaces ... he is surrounded on every side by this tough, safe, clean material which human thought has created ... [W]e shall see growing up around us a new, brighter cleaner and more beautiful world, an environment not subject to the haphazard distribution of nations' resources but built to order, the perfect expression of the new spirit of planned scientific control, the Plastics Age.
>
> (Yarsley and Couzens 1941: 149–52)

If you look up from this page, you will be surrounded by plastic. From the latex paint on the walls, to the carpeting, tiling or varnish on the floor, the chair you sit in, your shoes, watch and cell phone, right down to the elastic in your underpants, you are a Plastic (Wo)Man. Today, over 310 million tonnes of plastic are produced each year, accounting for around 8 per cent of the world's annual oil production (Andrady and Neal 2009: 1977; Thompson *et al.* 2009: 1973). Both figures are increasing annually.

The main difference between Yarsley and Couzens' futurist imaginings and the current state of plastics is that the Plastics Age is not occurring within a 'spirit of planned scientific control'. A generation into the Plastics Age, the longevity, durability and promiscuity of plastic chemicals has led to a new, poorly understood genre of pollution. The latest report from the Centers for Disease Control and Prevention (CDC) in the United States found that American bodies contained BPA, flame retardants, phthalates and polybrominated diphenyl ethers (PBDEs) – all chemicals that leach from plastics (CDCP 2009). In one form or another, and in startlingly high quantities, plastics can be found throughout numerous environments and bodies. Like miasmas, plastics are pervasive and their dangers are latent in everyday landscapes and objects.

Plastics are rarely just made of 'plastic' or polymers. Pure polyvinyl chloride (PVC) – the plastic used in most shower curtains, for example – is a white, brittle solid that does not possess the supple, mould-resistant, flower-patterned qualities of a shower curtain. To make many plastics more versatile, flexible, flame retardant, purple or countless other qualities, chemicals called plasticizers or plastic additives such as bis(2-ethylhexyl) phthalate (DEHP), BPA or polychlorinated biphenyls (PCB), must be included. Plasticizers are not

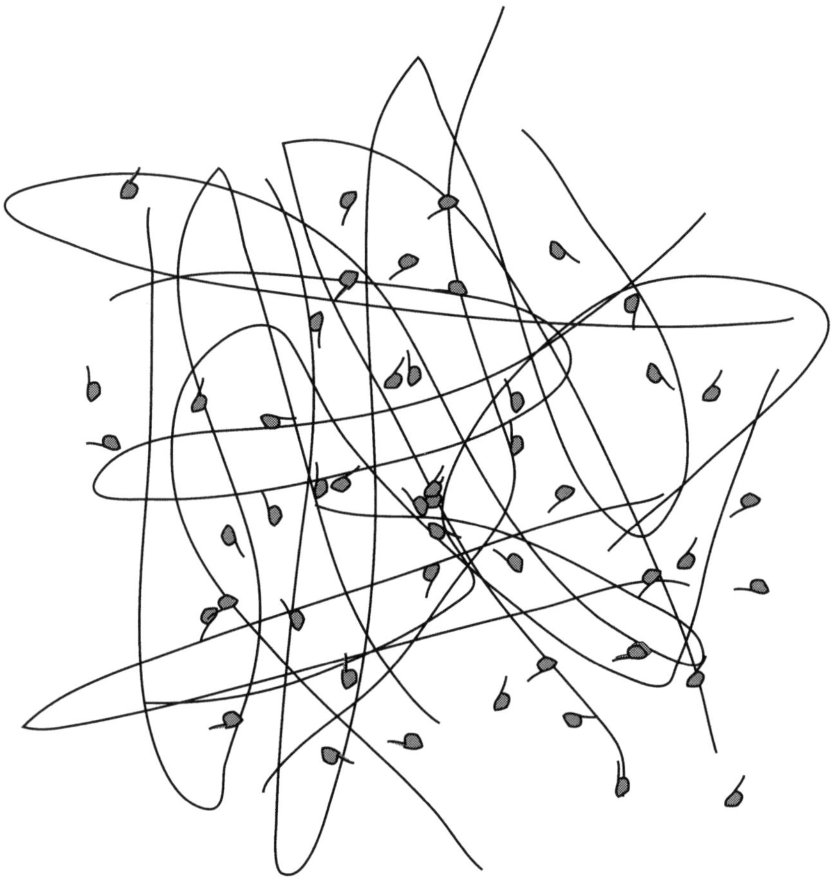

Figure 8.1 Long polymer strands make up plastics, while plasticizers nestle among the polymer strands, unbound, and can leave their host through off-gassing or leaching

chemically bound to the polymer chains, and thus can leave their hosts relatively easily (see Figure 8.1). A recent study found that 'Almost all commercially available plastic products … sampled, independent of the type of resin, product, or retail source, leached chemicals' (Yang *et al.* 2011). This ubiquity of plastics and promiscuity of plasticizers are the reasons why plastic chemicals accumulate in our bodies in such high quantities.

Plasticizers have a unique architecture that accounts for their mode of pollution. They are shaped like hormones. As such, they are classified as endocrine disruptors because they participate in the body's endocrine, or hormone, system rather than acting like foreign trespassers. Hormones and endocrine disruptors travel through the body until they encounter a receptor on a cell with a shape that complements their own. The hormone and receptor fit together like a lock and key (see Figure 8.2). When the two bind, the

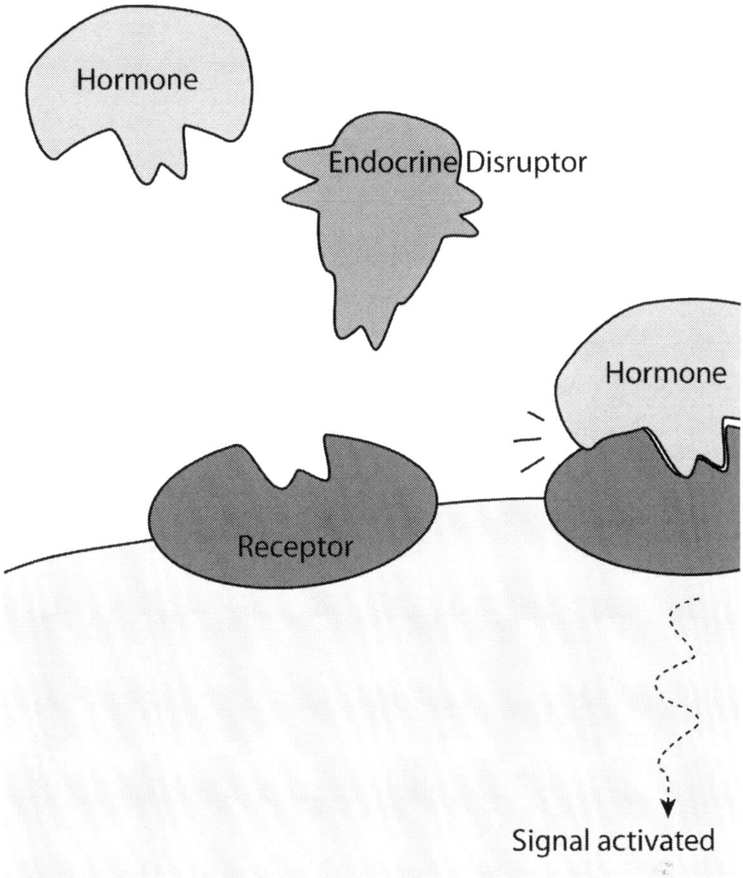

Figure 8.2 Endocrine receptors accept cells that complement their shape. Both hormones and endocrine disruptors can have similar shapes. On the right, a hormone and receptor have bound together and activated the receptor to signal DNA to begin work

activated receptor signals the DNA in the cell to get to work. Some of this work includes expressing genes, developing tissue and making proteins. This may result in nothing notable – a gene that expresses itself out of turn may just create extra or malformed, harmless proteins. On the other hand, various plasticizers have been correlated with infertility, recurrent miscarriages, feminization of male foetuses, early-onset puberty, obesity, diabetes, reduced brain development, cancer and neurological disorders such as early onset senility in adults and reduced brain development in children (Grun and Blumberg 2007: 8; Halden 2010: 179–94; Thompson *et al.* 2009).

The lock and key model of the endocrine system is not straightforward. There are several keys for each lock, each key can open several locks, each receptor has several locks and when a lock opens it can do several things. To continue with the metaphor, the sorts of things that unlocked locks can do – such as change the shape of sperm or influence the development of breasts – are also caused by completely different sets of locks and keys in other neighbourhoods. Some plasticizers mimic keys, some block locks, and some stop or increase the production of keys. The exponential complexity of these systems is not amenable to models based on the isolated pathways of discrete germs, but is better described as a web of influence. Sorting out singular effects and influences is exceedingly difficult, especially when the body's own hormonal effects are not distinguishable from what endocrine disruptors do. A body is never free of hormones, so testing the effects of an endocrine disruptor is always tied up with the effects of hormones that are already present. Moreover, since every body has multiple plasticizers in it at all times, they create what is called 'the cocktail effect'. Chemicals (and hormones) 'can react additively, multiplicatively or antagonistically' with one another. Disentangling what chemical caused which reaction becomes impossible because the chemicals influence one another (Meeker *et al.* 2009: 2108; Rider *et al.* 2008). Thus, a plasticizer under a microscope in a laboratory is something completely different from an endocrine disruptor in a body. The way in which they operate and their harmful effects are not separate from their environment.

Like Griscom's study of 'bad air', plasticizers affect every sub-system in the body, from the heart and its vascular system to the lungs and the respiratory system. For example, oestrogens and the endocrine disruptors that mimic them play a role in the development of sexual organs in women, the maturation of sperm and fertility in men, the maintenance of the skeletal system in both sexes, regulation of the menstrual cycle and pregnancy in women and the sex drive in both sexes. They help maintain memory functions, influence fat stores, support lung and heart function and promote mental health, and they influence the regulation of metabolism, protein synthesis, blood coagulation and the immune system, as well as salt and water retention. Oestrogens do all this in collaboration with other hormones (Nelson and Bulun 2001: S116–24). The effects of endocrine disruptors are not discrete; they operate more as non-specific influences.

Like Griscom's chilly and overheated gentlemen, different people respond to the same plasticizers in different ways. Endocrine disruptors have the greatest effects at the lowest doses, breaking the age-old toxicological rule that 'the danger is in the dose' (Vandenberg *et al.* 2009). The timing of the dose, the gender and age of the person, and the overall sensitivity and state of the ever-shifting endocrine system determine such relationships more than the dose itself. A specific dose of BPA to a female foetus will influence her differently than the same dose to an adult male, or even a male foetus twin (Hunt *et al.* 2003; Vandenberg *et al.* 2009). Furthermore, the effects of a dose may manifest in the person as a child, or during puberty, pregnancy or even menopause. If a female foetus is exposed, endocrine disruptors may influence her own gametes and thereby her future offspring, making effects intergenerational (Hunt *et al.* 2003: 555). This acute latency between the timing of exposure and its potential effect, compounded by the gender and state of the endocrine system at the time, make a direct relationship between exposure and effect impossible – perhaps even inadvisable – to isolate. Thus, due to the complexity of the system and the way plasticizers participate in it, the relationship between pollutant and body might best be described as a complex of influences and systems of causality.

Linear causality is not the only problem endocrine disruptors pose for the particle model: they also make it difficult to define harm. Some of the correlated effects of endocrine disruptors – such as cancer, delayed brain development and early onset dementia – readily fit into categories of harm. Yet others – such as obesity, the feminization of male foetuses and sexual differentiation – are challenging to categorize. Environmental historian Nancy Langston has undertaken work on diethylstilbestrol (DES), a synthetic endocrine disruptor given to many pregnant women in the 1930s, which has effects similar to many plasticizers. In her study, she cautions that:

> Understanding the potential for synthetic chemicals to cause birth defects requires more than medical research alone, for defining 'normal' and 'abnormal' is a social decision, not just a medical decision. When developmental changes in sexual characteristics have been linked to pollutants, distinguishing normal from abnormal is even more problematic.
>
> (Langston 2010: 140)

Is feminization of male foetuses abnormal, or even pathological? Is it a form of harm? The lesbian, gay, bisexual, transgender and queer (LGBTQ) community has argued that it is not. So, too, has the chemical industry. After research in 1995 found that low doses of BPA altered foetal brain tissue, 'a representative from the American Plastics Council pointed out that it was not possible to say whether this was a good or a bad thing for fetal brains' (Langston 2010: 146). Changes in gene expression and their effects can be difficult to define as harm. This does not necessarily mean endocrine disruptors fit neatly into miasmic categories of disease and health along lines of balance or imbalance, but nor do they fit rubrics of disease based on the presence or absence of pathogens or pollutants – especially given that plasticizers are always already present in all bodies.

The geography of plasticizers is both like and unlike miasma. Locally – that is, bodily – they are omnipresent. In an experiment where families ate food that was not processed or wrapped in plastic, individuals reduced their BPA burden by 60 per cent. This seems like a large amount until you consider that BPA is water soluble, and leaves the body in six to twelve hours, meaning the families were being continuously exposed to plasticizers through unknown pathways (Staples *et al.* 1998; Rudel *et al.* 2011). Plastics off-gas in homes, offices, manufacturing sites and outdoors. This lurking ubiquity is miasmic.

Yet in other aspects – such as their constant presence both inside and outside of bodies – plasticizers that act as endocrine disruptors are not like miasmas.

Miasma is local, while plastic pollution is global. Every winter, a high-pressure air system sweeps from the east coast of North America across the Atlantic Ocean and into Greenland. At the same time, air originating in mainland Europe is pushed into the High Arctic:

> In a matter of days or weeks, chemicals that originated in the cities of North America and Europe are contaminating the Arctic's air. When they reach the cold air, they condense and drop into the ocean or onto the frozen ground, where they are absorbed by plants, then animals [then people].
>
> (Cone 2005: 165)

This is one reason why indigenous Greenlanders are some of the most contaminated people on the planet. Moreover, some plasticizers – such as those used in our PVC shower curtain, mentioned earlier – are persistent organic pollutants (POPs), which are classified by their lack of discernible half-life, their ability to bioaccumulate, high levels of toxicity and the tendency to travel long distances. Over the next few thousand years, POPs will concentrate in the North, and many will stay there. This longevity also sets plasticizers apart from miasmas.

The influence model and action

The everyday, everywhere, miasmic geography of plasticizers is why avoiding plastic containers or banning BPA in baby bottles does not solve the problem of body burdens. As the experiment with families who avoided plastic containers attests, plasticizers are everywhere, inside and outside of plastics, inside and outside of bodies, and they are being produced in increasing quantities (Thompson *et al.* 2009). If BPA is not in the baby bottle, it is on the shopping receipt (Liao and Kannan 2011). Even if phthalates are banned outright, they do not degrade, and thus continue to circulate in the environment and in bodies. These mistaken expectations for successful intervention – avoidance and piecemeal bans for a pollutant that is omnipresent – is why turning to the influence model can shed new light on how to deal with and intervene in bodily plastic pollution. The way a problem is defined also defines the types of solutions that are considered viable and effective.

I am not the only researcher to find miasmas and the influence model useful for describing present-day ill health. In addition to its current use in some forms of homeopathy, in 1965 Bernard Bloom, an employee of the National Institute of Public Health, wrote an article calling for practitioners in mental health to consider the miasma model of disease over the 'medical model', a variation on what I have been calling the particle model (Bloom

1965; Tesh 1995). Both Bloom and I frame our arguments in terms of miasma because of the potential such a model has for action. Bloom argues that: 'The biological model contends that etiology and treatment are disease-specific. Accordingly, the establishment of a diagnosis is generally necessary before appropriate treatment can be instituted.' The miasma model, however, is 'a theory aimed at the prevention of disease', and thus allows for a wider and more immediate range of treatments and actions (Bloom 1965: 335). For Bloom, the miasma theory describes the dispersed causes of disease, and focuses on prevention and 'arranging optimal conditions for the patient to help himself' within a community, a solution he finds more effective than isolating diagnoses and assigning treatments (Bloom 1965: 335). His article ends with an extensive list of how nineteenth-century sanitarians reduced illness and mortality through preventative and community interventions.

Thinking of endocrine distributors through the logic of miasmas is useful for similar reasons: even while the mechanisms of action and harm of plastic pollution remain mysterious, mitigation is possible. This is not to say that we should import the interventions of eighteenth- and nineteenth-century sanitarians (which included dumping sewage into waterways), but that we should use an influence model to apprehend plastic pollution as matters of concern – entities with 'no clear boundaries, no well-defined essences, no sharp separation between their own hard kernel and their environment' – rather than attempting to define them as uncertain matters of fact, 'with their improbable cortege of incontestable knowledge, invisible scientists, predictable impacts, calculated risks, and unanticipated consequences' (Latour 2004: 24, 27). The influence model is particularly appropriate for policy makers, activists and those working to prevent bodily harm from plastics – particularly in the United States, where state and federal agencies consistently cite a lack of certainty, by which they mean clear linear causal data, as a reason to allow BPA and other plasticizers to remain listed as 'safe' substances (FDA 2012).

An influence model prioritizes the amorphous, contingent architecture and agency of plastic pollution as well as its pervasive geography, and judges proposed solutions to plastic pollution accordingly. For example, instead of targeting individual chemicals for bans such as the recent Canadian and American bans of BPA in baby bottles, all plasticizers – indeed, all chemicals used in consumer products – that act as endocrine disruptors should be heavily regulated or banned. This is the approach that the European Union (EU) has taken with its Registration, Evaluation, Authorisation, and Restriction of Chemicals (REACH) programme since 2007, where all chemicals and goods made with chemicals in the EU have to be registered and evaluated as 'safe' before they are permitted for use (Warhurst 2005). The legislation is so far-reaching that it has to specify that water is exempt (Warhurst 2005: 4). If plastics are everywhere – and some of them are permanent – these large-scale interventions make more sense than avoidance tactics. Broad-brush universal frameworks are better for regulating miasmas rather than piecemeal legislation that focuses on discrete chemicals one at a time.

Another advantage of using the influence model is that it reintroduces patterns within populations as a primary form of evidence. This has ramifications for findings that Northern indigenous populations are the most polluted people on the planet, and that, because of the condition of their endocrine systems, foetuses, children and women are the most vulnerable to endocrine disruptors. Though miasmas and plasticizers are ubiquitous, they have uneven concentrations and burdens of harm. Rather than focusing on presence alone, we can turn our attention to sources and topographies of harm and the types of interventions they demand. As in Griscom's time, the politics of concentration and where these concentrations originate can be reintroduced as part of the logic of pollution and its mitigation. So far, environmental justice has been under-utilized to describe the plastic pollution problem.

The influence model and its mysterious mechanics also legitimize the precautionary principle (e.g. see Marchant 2003). While there are various manifestations and interpretations of the precautionary principle, its main precept is that, if an action, policy or material may cause harm to the public or environment, but these consequences are uncertain or lack scientific consensus, either the action, policy or material should be abandoned or, in some formulations, the burden of proof that harm will not result falls on those taking the action or creating the substance (Marchant 2003; Luján and Todt 2012). REACH, mentioned above, uses the precautionary approach. It is a system that regulates 'mysterious' miasmic influences.

Finally, one of the greatest boons of the miasma model is that the precautionary approach privileges actions that produce results over those that do not, regardless of whether the mechanics of how and why they work are known. In 1854, John Snow convinced the Board of Guardians of St James' parish to remove the Broad Street pump handle during a cholera epidemic. Following this action, the epidemic died out (Johnson 2006). Snow's work is often cited as an early insight into germ theory, but it was carried out within an influence model of harm, where various interventions were attempted in the face of large-scale health crises, regardless of whether the mechanism for harm was known. Though the Board of Guardians rejected Snow's theory of water-borne illness, they still removed the pump handle. This historical moment, so often retrospectively appropriated by germ theorists, serves as a symbol of what a model of harm open to uncertainty, myriad influences and precaution can do for actions against miasmic phenomenon such as global plastic pollution.

This is not to say that miasmas and plastic pollution are synonymous. Miasmas are local, while plastic pollution is global. Miasmas often came from 'natural' sources such as swamps and manure, while plastic pollution is a synthetic chemical tied into an industrial economy dominated by multinational corporations. The influence model does not address the power dynamics, the extreme longevity or the industrial-consumer sources characteristic of plastic pollution that must be taken into account when attempting to mitigate the effects of plastic pollution. Instead, miasma provides a material framework for understanding how plastic pollution works in the environments and bodies of which it has become an everyday part. Miasmas provide a model, 'a tool for investigation, giving the user the potential to learn about the world' (Morgan and Morrison 1999) of plastic pollution in material terms so that solutions can reflect our increasingly plastic planet.

References

Abbott, A.C. (1899) *The Hygiene of Transmissible Diseases: Their Causation, Modes of Dissemination, and Methods of Prevention*, Philadelphia: W.B. Saunders.

Andrady, A.L. and Neal, M.A. (2009) 'Applications and Societal Benefits of Plastics', *Philosophical Transactions of the Royal Society B: Biological Sciences* 364(1526): 1977–84.

Bloom, B.L. (1965) 'The "Medical Model", Miasma Theory, and Community Mental Health', *Community Mental Health Journal* 1(4): 333–38.

Boyle, R. (1674) *Suspicions About Some Hidden Qualities of the Air*, London: William Godbid.

Bushnik, T., Haines, D., Levallois, P., Levesque, J., van Oostdam, J. and Viau, C. (2010) *Lead and Bisphenol-A Concentrations in the Canadian Population*, Ottawa: Statistics Canada.

Bynum, W.F. and Porter, R. (1993) *Companion Encyclopedia of the History of Medicine*, New York: Routledge.

Centers for Disease Control and Prevention (CDCP) (2009) *Fourth National Report on Human Exposure to Environmental Chemicals. Department of Health and Human Services*, Washington, DC: CDCP.

Chadwick, E. (1846) 'Metropolitan Sewage Committee Proceedings', in *Parliamentary Papers*, London: HM Government Printer.

Cipolla, C.M. (1992) *Miasmas and Disease: Public Health and the Environment in the Pre-industrial Age*, New Haven, CT: Yale University Press.

Cone, M. (2005) *Silent Snow: The Slow Poisoning of the Arctic*, New York: Grove Press.

Elliot, G.F. (1870) 'The Germ-theory', *The British Medical Journal* 1(489): 488–89.

FDA (Food and Drug Administration) (2012) *Re: FDMS Docket No. FDA-2008-P-0577-0001YCP*, Washington, DC: Department of Health and Human Services.

Federal Judicial Center (2011) *Reference Manual on Scientific Evidence*, Washington, DC: National Academy Press.

Griscom, J.H. (1854) *The Uses and Abuses of Air: Showing Its Influence in Sustaining Life, and Producing Disease; with Remarks on the Ventilation of Houses, and the Best Methods of Securing a Pure and Wholesome Atmosphere Inside of Dwellings, Churches, Courtrooms, Workshops, and Buildings of All Kinds*, New York: Redfield.

Grun, F. and Blumberg, B. (2007) 'Perturbed Nuclear Receptor Signaling by Environmental Obesogens as Emerging Factors in the Obesity Crisis', *Reviews in Endocrine & Metabolic Disorders* 8: 161–71.

Halden, R.U. (2010) 'Plastics and Health Risks', *Annual Review of Public Health* 31: 179–94.

Hill, H.W. (1916) *The New Public Health*, New York: Macmillan.

Hunt, P.A., Koehler, K.E., Susiarjo, M., Hodges, C.A., Ilagan, A., Voigt, R.C., Thomas, S., Thomas, B.F. and Hassold, T.J. (2003) 'Bisphenol-A Exposure Causes Meiotic Aneuploidy in the Female Mouse', *Current Biology* 13: 546–53.

Johnson, S. (2006) *The Ghost Map: The Story of London's Most Terrifying Epidemic – and How it Changed Science, Cities, and the Modern World*, New York: Riverhead Books.

Kern, S. (2004) *A Cultural History of Causality: Science, Murder Novels, and Systems of Thought*, Princeton, NJ: Princeton University Press.

Koch, R. (1901) 'The Combating of Tuberculosis', *Popular Science Monthly* 59: 461–74.

——(1961 [1882]) 'The Etiology of Tuberculosis', in T.D. Brock (ed.) *Milestones in Microbiology*, Englewood Cliffs, NJ: Prentice-Hall.

Langston, N. (2010) *Toxic Bodies: Hormone Disruptors and the Legacy of DES*, New Haven, CT: Yale University Press.

Latour, B. (1988) *The Pasteurization of France*, Cambridge, MA: Harvard University Press.

——(2004) *Politics of Nature: How to Bring the Sciences into Democracy*, Cambridge, MA: Harvard University Press.

Liao, C. and Kannan, K. (2011) 'High Levels of Bisphenol-A in Paper Currencies from Several Countries, and Implications for Dermal Exposure', *Environmental Science & Technology* 45(16): 6761–68.

Luján, J.L. and Todt, O. (2012) 'Precaution: A Taxonomy', *Social Studies of Science* 42(1): 143–57.

Marchant, G.E. (2003) 'From General Policy to Legal Rule: Aspirations and Limitations of the Precautionary Principle', *Environmental Health Perspectives* 111(14): 1799–803.

Meeker, J.D., Sathyanarayana, S. and Swan, S.H. (2009) 'Phthalates and Other Additives in Plastics: Human Exposure and Associated Health Outcomes', *Philosophical Transactions of the Royal Society B: Biological Sciences* 364(1526): 2097–113.

Morgan, M.S. and Morrison, M. (1999) *Models as Mediators: Perspectives on Natural and Social Sciences*, Cambridge: Cambridge University Press.

n.a. (1892) 'Waste Acid as a Pollutant', *Pal Mal Gazette* (London).

Nash, L.L. (2006) *Inescapable Ecologies: A History of Environment, Disease, and Knowledge*, Berkeley, CA: University of California Press.

Nelson, L.R. and Bulun, S.E. (2001) 'Estrogen Production and Action', *Journal of the American Academy of Dermatology* 45(3) (Supplement): S116–24.

Oxford University Press (2012) *Oxford English Dictionary*, Oxford: Oxford University Press.

Rider, C.V., Furr, J., Wilson, V.S. and Gray, L.E. Jr (2008) 'A Mixture of Seven Anti-androgens Induces Reproductive Malformations in Rats', *International Journal of Andrology* 31(2): 249–62.

Rudel, R.A., Gray, J.M., Engel, C.L., Rawsthorne, T.W., Dodson, R.E., Ackerman, J.M., Rizzo, J., Nudelman, J.L. and Green Brody, J. (2011) 'Food Packaging and Bisphenol-A and Bis(2-ethyhexyl) Phthalate Exposure: Findings from a Dietary Intervention', *Environmental Health Perspectives* 119(7): 914–20.

Staples, C.A., Dome, P.B., Klecka, G.M., Oblock, S.T. and Harris, L.R. (1998) 'A Review of the Environmental Fate, Effects, and Exposures of Bisphenol-A', *Chemosphere* 36(10): 2149–73.

Tesh, S.N. (1995) 'Miasma and "Social Factors" in Disease Causality: Lessons from the Nineteenth Century', *Journal of Health Politics, Policy and Law* 20(4): 1001–24.

Thompson, R.C., Moore, C.J., vom Saal, F.S. and Swan, S.H. (2009) 'Plastics, the Environment and Human Health: Current Consensus and Future Trends', *Philosophical Transactions of the Royal Society B: Biological Sciences* 364(1526): 2153–66.

Thornton, J. (2000) *Pandora's Poison: Chlorine, Health, and a New Environmental Strategy*, Cambridge, MA: MIT Press.

Vandenberg, L.N., Maffini, M.V., Sonnenschein, C., Rubin, B.S. and Soto, A.M. (2009) 'Bisphenol-A and the Great Divide: A Review of Controversies in the Field of Endocrine Disruption', *Endocrine Reviews* 30(1): 75–95.

Warhurst, M. (2005) *A Brief Introduction to the European Commission's Regulatory Proposal on Registration, Authorisation and Evaluation of Chemicals (REACH) L. C.f.S. Production*, Lowell, MA: Lowell Center for Sustainable Production, University of Massachusetts.

Yang, C.Z., Yaniger, S.I., Jordan, V.C., Klein, D.J. and Bittner, G.D. (2011) 'Most Plastic Products Release Estrogenic Chemicals: A Potential Health Problem that can be Solved', *Environmental Health Perspectives* 119(7): 989–96.

Yarsley, V.E. and Couzens, E.G. (1941) *Plastics*, Middlesex: Pelican.

9 Plastics, environment and health

Richard C. Thompson

Many current applications of plastics and their associated benefits follow those outlined in the 1940s by Yarsley and Couzens (1945). Their account of the benefits that plastics would bring to a person born 70 years ago, at the beginning of our 'plastic age', contained much optimism:

> It's a world free from moth and rust and full of colour, a world largely built up of synthetic materials made from the most universally distributed substances, a world in which nations are more and more independent of localised naturalised resources, a world in which man, like a magician, makes what he wants for almost every need out of what is beneath and around him.
>
> (Yarsley and Couzens 1945: 158)

The durability of plastics and their potential for diverse applications, including their widespread use in disposable items, was anticipated; however, the problems associated with waste management and accumulation of debris in the environment were not. In fact, the predictions were 'how much brighter and cleaner a world [it would be] than that which preceded this plastic age' (Yarsley and Couzens 1945: 58).

This chapter synthesizes current understandings of the benefits and concerns surrounding the use of plastics and addresses challenges, opportunities and priorities for the future. Central to this challenge will be finding ways to maximize the benefits that plastics can bring to society and the environment while at the same time minimizing any associated impacts. Many of the issues summarized here are discussed in more detail in the Royal Society's special theme issue, 'Plastics, the Environment and Human Health' (Thompson *et al.* 2009a, and summary papers 2009b and 2009c therein), and some of the solutions outlined are based around those described in a recent publication from the United Nations Environment Programme (UNEP), *Marine Debris as a Global Environmental Problem: Introducing a Solutions Based Framework Focused on Plastic* (GEF–STAP 2011).

Plastics are inexpensive, strong, lightweight and durable materials with high thermal and electrical insulation properties. The diversity of polymers and the versatility of their properties are utilized to make products with a wide variety of medical and technological advances, energy savings and numerous other societal benefits (Andrady and Neal 2009). On a global scale, plastic items are an essential part of daily life: in transport, telecommunications, clothing, footwear and as packaging materials, facilitating the distribution of a wide range of food, drink and other goods. There is considerable potential for new applications of plastics in the future – for example, as novel medical applications, in the generation of renewable energy and by reducing energy used in transport (Andrady and Neal 2009). As a consequence, the production of plastics has increased dramatically over the last 60 years, from around 0.5 million tonnes in 1950 to over 265 million tonnes today (PlasticsEurope 2011). Plastic production continues to grow by about 9 per cent annually, and the developed countries of Europe, Northern America and Japan account for about 60 per cent of global production and have the highest plastics consumption per capita of about 100–130 kg each year (PlasticsEurope 2008). The demand and consumption of plastics in developing countries on all continents is growing rapidly, driving a shift in production and conversion of plastics from developed to developing countries. The highest potential for growth is in the rapidly developing countries of Asia. Current consumption of plastic – like production – shows an exponential increase.

Environmental consequences of plastic debris

The majority of work describing the environmental consequences of plastic debris is based on marine settings. Plastic debris causes aesthetic problems, and also presents a hazard to maritime activities, including fishing and tourism (Gregory 2009; Moore 2008). In terms of larger debris, a particular concern is the accumulation of abandoned, lost or otherwise discarded fishing gear (ALDFG) from at-sea disposal, including fishing nets that continue to catch fish long after they have become marine debris. Plastics-based ALDFG can threaten marine habitats and fish stocks, and is also a concern for human health (Macfadyen *et al.* 2009). Floating plastic debris can rapidly become colonized by marine organisms, and, since it can persist at the sea surface, it may facilitate the transport of non-native or 'alien' species (Barnes 2002; Barnes *et al.* 2009; Gregory 2009). However, the problems attracting the most public and media attention are those resulting in ingestion and entanglement by wildlife. A recent review identified 373 reports of encounters between organisms and marine debris. Most of these reports related to entanglement in or ingestion of items of plastic debris (SCBD–STAP–GEF 2012), representing a 40 per cent increase on that reported in a previous review (Laist 1997). When considering the types of debris reported in relation to the categories of impact being addressed, it was clear that plastic items were by far most frequently documented, representing 76.5 per cent of all reports.

The most visible types of plastic debris are large derelict fishing gear, bottles, bags and other consumer products; however, much of the debris collected on shorelines and during survey trawls now consists of tiny particles, or 'microplastic' (Browne *et al.* 2011; Claessens *et al.* 2011; Collignon *et al.* 2012; Goldstein *et al.* 2012; Hidalgo-Ruz *et al.* 2012; Martins and Sobral 2011; Thompson *et al.* 2004). The term 'microplastic' was first used by Thompson and colleagues in 2004 to describe truly microscopic fragments, some of which were around 20 micrometres in diameter. Since then, the definition has been broadened to include small pieces or fragments less than 5 mm in diameter (Arthur *et al.* 2009). A horizon scan of global conservation issues recently identified microplastic as one of the top global emerging issues (Sutherland *et al.* 2010). Microplastic is formed by the physical, chemical and biological fragmentation of larger items, or from the direct release of small pieces of plastic, such as industrial spillage of pre-production pellets and powders, together with microscopic plastic particles that are used as abrasive scrubbers in domestic cleaning products (e.g. Fendall and Sewell 2009; Gouin *et al.* 2011) and industrial cleaning applications such as shot-blasting of ships and aircraft, and even fibres from domestic washing machines (Barnes *et al.* 2009; Browne *et al.* 2011). Plastic items fragment in the environment because of exposure to ultraviolet light and abrasion, causing smaller and smaller particles to form. Some plastic items are even designed to fragment into small particles, presumably so they are less conspicuous, but the resulting material does not necessarily biodegrade (Roy *et al.* 2011). Microplastics have accumulated in the water column, on the shoreline and in sub-tidal sediments (Browne *et al.* 2010; Thompson *et al.* 2004). Pieces as small as 2 micrometres have been identified (Ng and Obbard 2006), but due to limitations in sampling and analytical methods, the extent to which this type of debris has fragmented into nano-particle-size pieces is unknown. Microplastics are widely reported in trawls used to survey the abundance of plastic (Law *et al.* 2010; Thompson *et al.* 2004).

Laboratory experiments have shown that small pieces such as these can be ingested by a range of small marine organisms, including filter feeders, deposit feeders and detritivores (Thompson *et al.* 2004), while mussels were shown to retain plastic for over 48 days (Browne *et al.* 2008). A recent review indicates that around 8 per cent of all reported encounters in natural habitats between organisms and marine debris were with microplastics (SCBD–STAP–GEF 2012). Limited data exist on population-level exposure or consequences, but many of the birds surveyed by van Franeker *et al.* (2011) contain micro-plastic fragments, as do populations of the commercially important crustacean *Nephrops norvegicus*, where 83 per cent of individuals in the Clyde Sea contained microplastics. There is evidence that the abundance of microplastics is increasing (Thompson *et al.* 2004; Goldstein *et al.* 2012), and it is expected to increase further (Andrady 2011; Thompson *et al.* 2009b). Therefore, a number of important issues need to be investigated regarding the emissions, transport and fate, and physical and chemical effects of microplastics (Zarfl *et al.* 2011).

In addition to the physical problems associated with plastic debris, there is potential for plastic to transfer toxic substances to organisms if it is ingested. This concern has been raised with respect to microplastics because, due to their size, they are available to a wide range of organisms with various feeding strategies. Virgin plastic polymers are rarely used by themselves, and typically the polymer resins are mixed with additives to improve performance. These include inorganic fillers such as carbon and silica that reinforce the material, plasticizers to render the material pliable, thermal and ultraviolet stabilizers, flame retardants and colourings (see Meeker *et al.* 2009). Some additive chemicals are potentially toxic (Lithner *et al.* 2011), but there is controversy about the extent to which additives released from plastic products – such as phthalates and Bisphenol-A (BPA) – have adverse effects in animal or human populations. Additives of particular concern are phthalate plasticizers, BPA, brominated flame retardants and anti-microbial agents. BPA and phthalates are found in many mass-produced products including medical devices, food packaging, perfumes, cosmetics, toys, flooring materials, computers and CDs, and can represent a significant content of the plastic. For instance, phthalates can constitute 50 per cent of the total weight of polyvinyl chloride (PVC) (Oehlmann *et al.* 2009). There is evidence of the potential for these chemicals to be released to humans from plastic containers used for food and drink, plastic in medical applications and toys (Koch and Calafat 2009; Lang *et al.* 2008; Meeker *et al.* 2009; Talsness *et al.* 2009), and this has led to the introduction of legislation on human usage of items containing additives in some countries. Hence, these substances might potentially also be released if plastics containing them are ingested by marine organisms (Oehlmann *et al.* 2009; Teuten *et al.* 2009). While exposure pathways have not been determined, chemicals used in plastics – such as phthalates and flame retardants – have been found in fish, sea mammals, molluscs and other forms of marine life. This raises concerns about a potential for toxic effects. For example, there is evidence from laboratory studies about the adverse effects of BPA on a variety of aquatic organisms (Oehlmann *et al.* 2009; Talsness *et al.* 2009), and phthalates have also been shown to have adverse effects on aquatic organisms (Oehlmann *et al.* 2009). While a direct link between plastic debris and adverse effects on populations of marine organisms would be very difficult to demonstrate experimentally, if such effects were to occur there would be no simple way of reversing or remediating them due to the nature of debris accumulation in the environment (GESAMP 2010; Thompson *et al.* 2009b).

More work will be needed to establish the full environmental relevance of plastics in the transport of contaminants to organisms living in the natural environment, and the extent to which these chemicals could then be transported along food chains. However, there is already clear evidence that chemicals associated with plastic are potentially harmful to wildlife. Data collected using laboratory exposures show that phthalates and BPA affect reproduction in all studied animal groups, and impair development in crustaceans and amphibians (Oehlmann *et al.* 2009). Molluscs and amphibians appear to be particularly sensitive to these compounds, and biological effects have been observed at very low concentrations; in contrast, most effects in fish tend to occur at higher concentrations. Effects observed in the laboratory coincide with measured environmental concentrations, so there is a very real probability that these chemicals are affecting natural populations (Oehlmann *et al.* 2009). Chemicals such as phthalates and BPA can bioaccumulate in organisms, but there is much variability between species and individuals according to the type of plasticizer and experimental protocol. While these chemicals have adverse effects at environmentally relevant concentrations in laboratory studies, there is a need for further research to establish population-level effects in the natural environment, to establish the long-term effects of exposures (particularly due to exposure of embryos), to determine effects of exposure to contaminant mixtures, and to establish the role of plastics as sources (albeit not exclusive sources) of these contaminants (see Meeker *et al.* 2009 for discussion of sources and routes of exposure).

Concerns for human health

Turning to the adverse effects of plastic on the human population, there is a growing body of literature on potential health risks. A range of chemicals that are used in the manufacture of plastics are known to be toxic. Measuring concentrations of environmental contaminants in human tissue, or biomonitoring, has shown that chemicals used in the manufacture of plastics are present in the human population. Interpreting biomonitoring data is complex, and a key task is to put information in perspective with dose levels that are considered toxic, on the basis of experimental studies in laboratory animals (e.g. see Myers *et al.* 2009; Talsness *et al.* 2009). The biomonitoring information we have demonstrates that phthalates and BPA, as well as other additives in plastics and their metabolites, are present in the human population. There are differences according to geographic location and age, with greater concentrations of some of these chemicals in young children. Koch and Calafat (2009) show that, while mean/median exposures for the general population were below levels determined to be safe for daily exposure (USA, EPA reference dose, RfD; and European Union (EU) tolerable daily intake, TDI), the upper percentiles of Dibutyl phthalate (DBP) and di-2-ethylhexyl phthalate (DEHP) urinary metabolite concentrations show that for some people daily intake might be substantially higher than previously assumed, and could exceed estimated safe daily exposure levels. The toxicological consequences of such exposures – especially for susceptible sub-populations such as children and pregnant women – remain unclear, and warrant further investigation. However, there is evidence of associations between urinary concentrations of some phthalate metabolites and biological outcomes (Swan 2008). For example, an inverse relationship has been reported between the concentrations of DEHP metabolites in the mother's urine and anogenital distance, penile width and testicular descent in male offspring (Swan 2008). In adults, there is some evidence of a negative association between phthalate metabolites and semen quality, and between high exposures to phthalates (workers producing PVC flooring) and free testosterone levels. Moreover, recent work (Lang *et al.* 2008) has shown a significant relationship between urine levels of BPA and cardiovascular disease, Type 2 diabetes and abnormalities in liver enzymes. These data indicate that detrimental effects in the general population may be caused by chronic low-dose exposures (separately or in combination) and acute exposure to higher doses, but the full extent to which chemicals are transported to the human population by plastics is yet to be confirmed.

Despite the environmental concerns about some of the chemicals used in plastic manufacture, it is important to emphasize that evidence for effects in humans is limited, and there is a need for further research – particularly for longitudinal studies to examine temporal relationships with chemicals that leach out of plastics (Adibi *et al.* 2008). In addition, the traditional approach to studying toxicity of chemicals has been to focus only on exposure to individual chemicals in relation to disease or abnormalities. However, because of the complex integrated nature of the endocrine system, it is critical that future studies focus on mixtures of chemicals to which people are exposed when they use common household products. For example, 80 per cent of babies in a study conducted in the United States were exposed to measurable levels of at least nine different phthalate metabolites (Sathyanarayana *et al.* 2008), and the health impacts of the cumulative exposure to these chemicals need to be determined. An initial attempt at examining more than one phthalate as a contributor to abnormal genital development in babies has shown the importance of this approach (Swan 2008).

Examining the relationship between plastics additives and adverse human effects presents a number of challenges. In particular, the changing patterns of production and use of both plastics and the additives they contain, as well as the confidential nature of industrial specifications, make exposure assessment particularly difficult. Evolving technology, methodology and statistical approaches should help to disentangle these relationships. However, with most of the statistically significant hormone alterations that have been attributed to environmental and occupational exposures, the actual degree of hormone alteration has been considered sub-clinical, and more information is required on the biological mechanisms that may be affected by low-dose chronic exposures. Meanwhile, we should consider strategies to reduce the use of these chemicals in plastic manufacture and/or develop and test alternatives. This is the goal of the new field of green chemistry, which is based on the premise that development of chemicals for use in commerce should involve an interaction between biologists and chemists. Had this approach been in place 50 years ago, it would probably have prevented the development of chemicals that are recognized as likely endocrine disruptors (Anastas and Beach 2007; Thompson *et al.* 2009c).

Solutions: maximizing the benefits and reducing impacts

There is a considerable volume of scientific literature, together with publications from governments, nongovernmental organizations (NGOs), academia and industry, outlining our understanding of issues relating to plastics, the environment and human health (Thompson *et al.* 2009a). It is clear from this literature that there are unresolved knowledge gaps, uncertainties, and areas of disagreement and debate. A summary of existing knowledge and current uncertainty is given in Thompson *et al.* (2009c: table 1), in Zarfl *et al.* (2011) and in GEF–STAP (2011). While resolving uncertainties will clearly help us to refine solutions and prioritize, the author considers that there is broad agreement from all quarters – industry, policy, public, academia and NGOs – that a reduction in marine debris, particularly plastic debris, is a priority that requires action as a matter of urgency.

Some of the major sources of marine debris are well described, and include sewage- and run-off-related debris, materials from recreational/beach users, and materials lost or disposed of at sea from fishing activities (such as ALDFG) or shipping (Derraik 2002; OSPAR 2007; Thompson *et al.* 2009c; UNEP 2009). Debris originating from the land is either transported by storm water, via drains and rivers, towards the sea, or is blown into the sea (Macfadyen *et al.* 2009; Ryan *et al.* 2009). Extreme weather events, such as hurricanes and floods, are important point sources of marine debris from/to the sea (Thompson *et al.* 2005). Sea-based sources of debris represent additional, and in some regions substantial, sources of debris. The dumping of waste at sea is regulated by many agreements and conventions, and, while there are problems with enforcement, reductions in the amount of debris from ship-based activities have been reported in some regions. Two commonly used tools in reducing ship-based sources of marine debris are the availability of appropriate and convenient port reception facilities for waste from ships (Mouat *et al.* 2010; Thompson *et al.* 2009c) and educational materials (such as multi-language posters and video footage).

ALDFG has been recognized internationally as a major problem, and proposals for addressing it have been made at the level of the UN General Assembly (UNGA) and its specialized agencies and programmes, including the Food and Agriculture Organization of the UN (FAO), the UNEP and the International Maritime Organization (IMO). There have also been regional calls to address ALDFG. Initiatives to reduce ALDFG are crucial, and implementation principles are generally similar to measures addressing land-based sources of marine debris, such as discarded consumer goods and packaging (prevention, mitigation, removal and awareness-raising), but there are many important sector-specific issues.

There are, however, much broader causes and responsibilities, spanning production, use and disposal of the items that become marine debris. Furthermore, the origins of the problems and their solutions lie not only in coastal communities, but also far inland. They are rooted in production and consumption patterns – including the design and marketing of products without appropriate consideration for their environmental fate or their ability to be recycled in the locations where sold – as well as within inadequate waste-management practices and irresponsible behaviour. In addition, there can be considerable geographical separation between production, which is often centred in relatively developed economies, and consumption/disposal, which is global. Hence, the widely accepted proposition that the problem of marine debris is associated predominantly with poor management practices on land (Andrady and Neal 2009; PlasticsEurope 2010; UNEP 2005) needs to be expanded considerably. A broader approach, such as that summarized recently by UNEP (GEF–STAP 2011), is now gaining momentum among governments, NGOs and some industries (EU Director-General Environment 2011a, 2011b; Kershaw *et al.* 2011; UNEP 2009; US Commission on Ocean Policy 2004).

An essential part of this discussion is to recognize the advantages that plastic products bring to society and to the environment, both currently and in the future. Plastics have undoubted benefits in medical, educational and transport applications; they also have a key role to play in helping reduce humankind's footprint on the environment (PlasticsEurope 2008). Use of plastic components in automobiles and aircraft results in significant weight savings compared with metals. The new Boeing 787 aircraft will, for example, have a skin that is 100 per cent composite and an interior that is 50 per cent plastic, resulting in combined fuel savings of around 20 per cent (Andrady and Neal 2009). Single-use plastic packaging items are among the most common components of marine litter; however, such packaging has a key role in reducing wastage since even a relatively small use of packaging can extend the shelf life of perishable products, hence contributing to food safety. Since plastic packaging is lightweight, it can also achieve significant reductions in fuel usage (packaging in PET can achieve a 52 per cent saving over glass, for example) during transportation (Andrady and Neal 2009). It is this combined success that has led to global production of plastics, accounting for around 4 per cent of world oil production in the products themselves and a further 4 per cent in the energy required for this production. However, this success also results in the accumulation of end-of-life plastics that is being examined here. In order to generate solutions, an holistic framework is required, which aspires to optimize benefits and reduce impacts in order to harness the greatest potential that plastic products can offer humanity.

A key challenge in addressing the problems associated with plastic debris is broadening the range of available management measures beyond improvement in waste-management practices (EU Director-General Environment 2011b; UNEP 2009). At present, these are predominantly 'end-of-pipe' responses, rather than preventative measures. The most commonly used approaches vary regionally, but include educational notices about the problems of dumping and littering, improved reuse, recycling and recovery (under strictly controlled conditions), provision of litter bins on beaches, port reception for waste from ships, and extensive clean-up campaigns on shorelines and at sea. The plastics industry has supported end-of-life consumer education and recycling programmes as a solution (e.g. see Marine Debris Solutions n.d.), but such measures are more relevant to highly developed nations with economic resources and economies of scale to make the programmes cost-effective. In relation to ever-increasing global and regional trends in the quantity of plastics waste being produced, it becomes clear that a paradigm shift is required in the way we address this global problem.

From a life-cycle perspective, linear use of resource from production – and, in the case of single-use packaging, through a short-lived usage stage – to disposal is a central underlying cause of the accumulation of waste (Thompson *et al.* 2009c; WRAP 2006). Much of the current production and consumption lacks long-term sustainability because the amount of raw materials and our capacity to deal with waste are finite (Barnes *et al.* 2009; Thompson *et al.* 2009c). If there are manufactured products and associated packaging, there is a potential source of debris. Put simply, if we can reduce the quantity of plastic waste we produce at the same time as improving waste management options – for example, via recycling – we will maximize our potential to tackle the problems associated with accumulation of waste in the environment and in landfills. While trends in the quantity of debris parallel economic growth, the pathways that cause plastic debris to enter the environment are trans-boundary and global in nature, and it is this disconnect that necessitates new approaches.

Recognizing that marine debris is not merely a waste-management issue is fundamental to addressing the underlying causes of this debris. Solving the marine plastics debris problem through a life-cycle approach is therefore one of the potential testing grounds for the green economy concept: this will involve using fewer resources per unit of economic output, and reducing the environmental impact of any resources that are used or economic activities that are undertaken without compromising growth. Applied to plastics, this means promoting structural economic changes that would reduce plastics consumption, increase production of environmentally friendlier materials, increase recycling and reuse, promote investments in alternative conversion technologies and new materials and products, and support an enabling environment including capacity building, and new regulations and standards (Thompson *et al.* 2009c). Such benefits can only be realized by working in partnership with industry: the benefits of collaboration are recognized by both policy – for example, the Congressionally mandated US Commission on Ocean Policy (2004) – and industry (APR 2011), and acknowledged within the EU (EU Director-General Environment 2011b).

While the problems of plastic debris have both trans-boundary and global sources and causes, the types and quantities of this debris, and their impacts, have strong regional components. Numerous relatively generic approaches have been identified to reduce the amount of debris produced, better to manage the waste that is produced and to remove from aquatic habitats the waste that has accumulated. Since packaging comprises a substantial proportion of the plastic items produced, and also makes up one of the major categories of plastics within marine debris, the following examples will focus on the production, use and disposal of single-use packaging. The three Rs – reduce, reuse, recycle – are widely advocated to reduce the quantities of waste, particularly the plastics packaging waste we generate. To be effective, we need to consider the interconnectivity between these three Rs in combination with each other and together with a fourth 'R' – redesign – including both molecular redesign via green chemistry approaches and product redesign with greater resource efficiency and environmental sustainability in mind as an emerging and potentially very important strategy. For items that cannot be designed for reuse or recycling, a fifth 'R' – energy recovery – can be considered. Hence, the three Rs become five: reduce, reuse, recycle, redesign and recover.

There are opportunities to *reduce* usage of raw material by product redesign and opportunities to *reuse* plastics – for example, in the transport of goods at an industrial (pallets, crates) and a domestic (reusable carrier bags) scale. However, there is often limited potential for wide-scale reuse of packaging because of the substantial back-haul distances and logistics involved in returning empty cartons to suppliers, especially in communities or regions with an under-developed infrastructure. Perhaps most importantly, there is now a strong evidence base to indicate the significant potential that lies in increasing our ability to effectively *recycle* end-of-life plastic products. Although thermoplastics have been recycled since the 1970s, the proportion of material recycled has increased substantially in some countries in recent years (APR 2011; PlasticsEurope 2008; WRAP 2006). Plastics can be designed to be inherently very recyclable, and there is considerable potential to turn end-of-life items back into new items. There is now strong evidence that significant potential lies in increasing our capacity to recycle end-of-life plastics; recycling can be economically attractive, and can reduce carbon dioxide emissions compared with use of new polymer (DEFRA 2007; WRAP 2006, 2008). However, it may be advantageous to introduce further economic incentives to encourage the redesign of plastic items to be both more reusable and recyclable, and to increase local and regional opportunities for recycling, in order to achieve broader geographic spread. Recognizing the potential value of end-of-life plastics as a raw material for new production not only reduces waste in the environment, but also incentivizes careful disposal as opposed to littering, reduces reliance on non-renewable oil and gas resources, and as a whole is likely to generate global environmental and economic benefits (Thompson *et al.* 2009c). While this indicates the potential, there is still much more that can be done to increase the spatial extent and increase the proportion of plastic items that are recycled. The recycling message is simple: both industry and society need to view end-of-life plastic as a raw material rather than waste. Greatest energy efficiency is achieved where recycling diverts the need for use of fossil fuels as raw materials – a good example is the recycling of polyethylene terephthalate (PET) bottles into new ones (closed-loop recycling) (Hopewell *et al.* 2009).

Historically, the main considerations for the design of plastic packaging have been getting goods safely to market and product marketing. These are, of course, important for food safety and for industry; however, there is an increasing urgency also to design plastic products – especially packaging – for material reduction, reuse and high end-of-life recyclability. Public support for recycling is high in some countries (57 per cent in the United Kingdom and 80 per cent in Australia), and consumers are keen to recycle (Hopewell *et al.* 2009); however, the small size and diversity of different symbols to describe a product's potential recyclability, together with uncertainties about whether a product will actually be recycled if collected, have the potential to hinder engagement (Hopewell *et al.* 2009; Thompson *et al.* 2009c). In addition, recycling requires significant investment in infrastructure for collection, transport, sorting and management of the recyclable items. While such infrastructure can be economically feasible in developed nations, it may not be feasible or cost-effective for developing countries. Hence, the need exists for a regionally centred framework (GEF–STAP 2011). From an industry perspective, molecular *redesign* of plastics (the fourth 'R') has become an emerging issue in green chemistry. In this context, green chemists aspire to design chemical products that are: (i) fully effective; (ii) yet have little or no toxicity or endocrine-disrupting activity; (iii) break down into innocuous substances if released into the environment after use; and/or (iv) are based upon renewable feedstocks, such as agricultural wastes. Such approaches should be considered within the design and life-cycle analysis of plastics – for example, to maximize the proportion of plastic products that are recycled (EU Director-General Environment 2011a, 2011b).

One of the fundamental factors limiting progress on the original three Rs (reduce, reuse and recycle) is that the design criteria used to develop new polymers and products seldom include specifications to enhance reusability, recyclability or recovery of plastic once it has been used. Typically, such assessments have only been made after products have entered the marketplace and been recognized as having unintended consequences, but such sentiments are also being echoed by the recycling sector, the trade association of which considers that 'the guiding principle of any packaging design must be fitness of purpose. Beyond this, designing to enhance recyclability should be in the forefront of design considerations' (APR 2011).

While industry and policy makers concur on the need to seek innovative solutions that go beyond end-of-pipe recycling (e.g. EU Director-General Environment 2011b; PlasticsEurope 2008), it is essential that this is achieved in collaboration. The dangers of working in isolation are already apparent from industry-centred responses such as the development of 'oxo-degradable' plastic products, which merely fragment at the end of their lifetime into numerous small but essentially non-degradable pieces, the environmental impact of which is not yet known (Roy *et al.* 2011). Degradable materials such as these also compromise recycling, and there are concerns about their efficacy (DEFRA 2010). From a different perspective, working in isolation can also lead to policy-centred responses, such as a blanket ban on plastic bags, rather than promoting the use of reusable bags, including those made of plastic.

After proper consideration of the preceding Rs, for plastic products that cannot be redesigned and plastic waste that cannot be reused or recycled, some of the energy content can be 'recovered' by incineration, and through approaches such as co-fuelling of kilns. This can be reasonably energy-efficient, but multiple tradeoffs have to be accounted for before such a decision is made (Thompson *et al.* 2009c), and, unless appropriate regulations are in place, combustion of plastics may result in release of chlorinated dioxins, furans and other persistent toxic compounds such as brominated dioxins. Due to these negative impacts, combustion of plastic debris could be a serious environmental issue in developing countries if end-of-life energy recovery is not considered by the plastic manufacturer when designing the plastic or product, or if the energy-recovery system is inadequately regulated. While energy recovery for certain types of plastic waste has benefits compared with disposal to landfill, energy recovery does not reduce the demand for raw material used in production. Hence, it is considered much less desirable and less energy-efficient than reuse or product recovery via recycling.

Specific examples of the Rs will only succeed if they are based on regional priorities, and oriented towards the needs and perspectives of the consumers/users in particular regions and nations. Solutions should then be identified through cooperation between industry, government and consumer, and consider all five Rs in a regionally relevant context. Potential actions to consider in this context encompass any or all parts of the supply and value chain, and the full life-cycle of the product, including: (i) educating users; (ii) collecting and removing debris from the environment; (iii) measures to reduce the production of waste or improve the design of the product itself; and (iv) considering extended producer responsibility (EPR) to achieve these goals. EPR is crucial, as it will help redistribute the burden of handling end-of-life plastic from governments and individuals who may be impacted by the waste, despite having no say in the product, to producers with interests that would then be more closely aligned with the problems encountered in product disposal.

Biopolymers, degradable and biodegradable polymer solutions

Degradable polymers have been advocated as an alternative to conventional oil-based plastics, and their production has increased considerably in recent decades. Before concluding our discussion of solutions to the problem of plastics waste and pollution, a brief consideration of the applications and limitations of these novel materials is worthwhile. Biopolymers – degradable and biodegradable polymers with comparable functionality to conventional plastics – can now be produced on an industrial scale; however, they are more expensive than conventional polymers and account for less than 1 per cent of plastics production (Song *et al.* 2009). Biopolymers differ from conventional polymers in that their feedstock is from renewable biomass rather than being oil based. They may be natural polymers (e.g. cellulose) or synthetic polymers made from biomass monomers (e.g. polylactic acid – PLA), or they could be synthetic polymers made from synthetic monomers derived from biomass (e.g. polythene derived from bioethanol) (WRAP 2009). They are often described as renewable polymers since the original biomass – for example, corn grown in agriculture – can be reproduced. The net carbon dioxide emissions may be less than that with conventional polymers, but they are not zero, since farming and pesticide production have carbon dioxide outputs (WRAP 2009). In addition, as a consequence of our rapidly increasing human population, it seems unlikely that there will be sufficient land to grow crops for food, let alone for substantial quantities of packaging to wrap it in. One solution is to recycle waste food into biopolymers; this has merit, but ultimately will be limited by the amount of waste food available.

Biopolymers that are designed to break down in an industrial composter are described as 'biodegradables', while those that are intended to degrade in a domestic composter are known as 'compostables'. These biodegradable materials are of benefit in specific applications – for example, the packaging of highly perishable goods where, regrettably, it can be necessary to dispose of perished unopened and unused products together with their packaging. Song *et al.* (2009) show experimentally that degradation of biodegradable, as opposed to compostable, polymers can be very slow in home composters (typically less than 5 per cent loss of biomass in 90 days). Degradation of these polymers in landfills is also likely to be slow, and may create unwanted methane emissions. Hence, the benefits of biopolymers are only realized if they are disposed of using an appropriate waste management system that utilizes their biodegradable features. Typically, this is achieved via industrial composting at 50°C for around 12 weeks to produce compost as a useful product.

Some biopolymers, such as PLA, are biodegradable but others, such as polythene derived from bioethanol, are not. In addition, degradable polymers (also called 'oxo-biodegradable', 'oxy-degradable' or 'UV-degradable') can be made from oil-based sources, and as a consequence are not biopolymers. These degradable materials are typically polyethylene, together with additives to accelerate the degradation. They are used in a range of applications, and are designed to break down under UV (ultraviolet) exposure and/or dry heat and mechanical stress, leaving small particles of plastic behind. They do not degrade effectively in landfill, and little is known about the timescale, extent or consequences of their degradation in the natural environment (Barnes *et al.* 2009; Teuten *et al.* 2009). Degradable polymers may also compromise the quality of recycled plastics if they enter the recycling scheme. As a consequence, use of degradable polymers is not advocated for primary retail packaging (WRAP 2009).

There is a popular misconception that degradable and biodegradable polymers offer solutions to the problems of plastic debris and the associated environmental hazards that result from littering. However, most of these materials are unlikely to degrade quickly in natural habitats, and there is concern that degradable, oil-based polymers could merely disintegrate into small pieces that are not in themselves any more degradable than conventional plastic (Barnes *et al.* 2009). So while biodegradable polymers offer some waste-management solutions, there are limitations to their use, as well as considerable misunderstanding among the general public about their application (WRAP 2007). To gain the maximum benefit from degradable, biodegradable and compostable materials, it is therefore essential to identify specific uses that offer clear advantages and to refine national and international standards (e.g. EN 13432, ASTM D6400-499) and associated product labelling to indicate appropriate usage and disposal methods.

Conclusion

Looking ahead, we are clearly not approaching the end of the 'plastic age' described by Yarsley and Couzens (1945) almost 70 years ago, and there is much that plastics can contribute to society. Andrady and Neal (2009) consider that the speed of technological change is increasing exponentially, such that life in 2030 will be unrecognizable compared with life today; plastics will play a significant role in this change, and have the potential to bring scientific and medical advances, alleviate suffering and help reduce humankind's environmental footprint on the planet (Andrady and Neal 2009). However, it is evident that our current approaches to the production, use and disposal of plastics are not sustainable, and present concerns for both wildlife and human health. We have considerable knowledge about many of the environmental hazards, and information on human health effects is growing; however, many concerns and uncertainties remain. There are solutions, but these can only be achieved by combined actions. There is a role for individuals, via appropriate use and disposal – particularly recycling; for industry, by adopting green chemistry and material reduction, and designing products for reuse and/or end-of-life recyclability; and for governments and policy makers by setting standards and targets, defining appropriate product labelling to inform and incentivize change, and funding relevant academic research and technological developments. These actions are overdue, and now need to be implemented with urgent effect; there are diverse environmental hazards associated with the accumulation of plastic waste and there are growing concerns about plastic's effects on human health, yet plastic production continues to grow at around 9 per cent per annum (PlasticsEurope 2008). As a consequence, the quantity of plastics produced in the first 10 years of the current century will approach the total that was produced in the entire twentieth century (Thompson *et al.* 2009b).

References

Adibi, J.J., Whyatt, R.M., Williams, P.L., Calafat, A.M., Camann, D., Herrick, R., *et al.* (2008) 'Characterization of Phthalate Exposure Among Pregnant Women Assessed by Repeat Air and Urine Samples', *Environmental Health Perspectives* 116: 467–73.

Anastas, P.T. and Beach, E.S. (2007) 'Green Chemistry: The Emergence of a Transformative Framework', *Green Chemistry Letters and Reviews* 1(1): 9–24.

Andrady, A.L. (2011) 'Microplastics in the Marine Environment', *Marine Pollution Bulletin* 62: 1596–605.

Andrady, A.L. and Neal, M.A. (2009) 'Applications and Societal Benefits of Plastics', *Philosophical Transactions of the Royal Society B* 364(1526): 1977–84.

Arthur, C., Baker, J. and Bamford, H. (2009) *Proceedings of the International Research Workshop on the Occurrence, Effects and Fate of Microplastic Marine Debris. September 9–11, 2008*, NOAA Technical Memorandum NOS-OR&R30.

Association of Postconsumer Plastic Recyclers (APR) (2011) *The Association of Postconsumer Plastic Recyclers Design for Recyclability Program*, Washington, DC: APR.

Barnes, D.K.A. (2002) 'Biodiversity: Invasions by Marine Life on Plastic Debris', *Nature* 416: 808–9.

Barnes, D.K.A., Galgani, F., Thompson, R.C. and Barlaz, M. (2009) 'Accumulation and Fragmentation of Plastic Debris in Global Environments', *Philosophical Transactions of the Royal Society B* 364(1526): 1985–98.

Brink, P.T., Lutchman, I., Bassi, S., Speck, S., Sheavly, S. *et al.* (2009) *Guidelines on the Use of Market-based Instruments to Address the Problem of Marine Litter*, Brussels: Institute for European Environmental Policy (IEEP).

Brinton, W.F. (2005) 'Characterization of Man-made Foreign Matter and its Presence in Multiple Size Fractions from Mixed Waste Composting', *Compost Science & Utilization* 13: 274–80.

Browne, M.A., Crump, P., Niven, S.J., Teuten, E., Tonkin, A. *et al.* (2011) 'Accumulation of Microplastic on Shorelines Worldwide: Sources and Sinks', *Environmental Science & Technology* 45: 9175–79.

Browne, M.A., Dissanayake, A., Galloway, T.S., Lowe, D.M. and Thompson, R.C. (2008) 'Ingested Microscopic Plastic Translocates to the Circulatory System of the Mussel, Mytilus Edulis (L.)', *Environmental Science and Technology* 42: 5026–31.

Browne, M.A., Galloway, T.S. and Thompson, R.C. (2010) 'Spatial Patterns of Plastic Debris Along Estuarine Shorelines', *Environmental Science & Technology* 44: 3404–9.

Claessens, M., de Meester, S., van Landuyt, L., de Clerck, K. and Janssen, C.R. (2011) 'Occurrence and Distribution of Microplastics in Marine Sediments Along the Belgian Coast', *Marine Pollution Bulletin* 62: 2199–204.

Collignon, A., Hecq, J.H., Galgani, F., Voisin, P., Collard, F. and Goffart, A. (2012) 'Neustonic Microplastic and Zooplankton in the North Western Mediterranean Sea', *Marine Pollution Bulletin* 64: 861–64.

DEFRA (Department of Environment Food and Rural Affairs) (2007) *Waste Strategy for England*, Norwich: DEFRA.

——(2010) *Assessing the Environmental Impacts of Oxo-degradable Plastics Across Their Life Cycle*, Norwich: DEFRA.

DEFRA, Enviros, Wilson, S. and Hannan, M. (2006) *Review of England's Waste Strategy, Environmental Report under the 'SEA' Directive*, Norwich: DEFRA.

Derraik, J.G.B. (2002) 'The Pollution of the Marine Environment by Plastic Debris: A Review', *Marine Pollution Bulletin* 44: 842–52.

EU Director-General Environment (2011a) *Plastic Waste in the Environment*, Brussels: European Union.

——(2011b) *Plastic Waste: Redesign and Biodegradability*, Science for Environment Policy Future Briefs, No. 1, Brussels: European Union.

Fendall, L.S. and Sewell, M.A. (2009) 'Contributing to Marine Pollution by Washing your Face: Microplastics in Facial Cleansers', *Marine Pollution Bulletin* 58: 1225–28.

GEF–STAP (Global Environmental Facility Scientific and Technical Advisory Panel) (2011) *Marine Debris as a Global Environmental Problem: Introducing a Solutions Based Framework Focused on Plastic*, Washington, DC: Global Environment Facility, www.thegef.org/gef/pubs/STAP/marine-debris-defining-global-environmental-challenge (accessed 20 August 2012).

GESAMP (2010) 'Proceedings of the GESAMP International Workshop on Plastic Particles as a Vector in Transporting Persistent, Bio-accumulating and Toxic Substances in the Oceans', in T. Bowmer and P.J. Kershaw (eds) *GESAMP Reports and Studies*, Geneva: UNESCO.

Goldstein, M., Rosenberg, M. and Cheng, L. (2012) 'Increased Oceanic Microplastic Debris Enhances Oviposition in an Endemic Pelagic Insect', *Biology Letters* 10.1098 (published online 9 May).

Gouin, T., Roche, N., Lohmann, R. and Hodges, G. (2011) 'A Thermodynamic Approach for Assessing the Environmental Exposure of Chemicals Absorbed to Microplastic', *Environmental Science & Technology* 45: 1466–72.

Gregory, M.R. (2009) 'Environmental Implications of Plastic Debris in Marine Settings: Entanglement, Ingestion, Smothering, Hangers-on, Hitch-hiking, and Alien Invasions', *Philosophical Transactions of the Royal Society B* 364(1526): 2013–26.

Harper, P.C. and Fowler, J.A. (1987) 'Plastic Pellets in New Zealand Storm-killed Prions (*Pachyptila* spp.), 1958–77', *Notornis* 34: 65–70.

Hidalgo-Ruz, V., Gutow, L., Thompson, R.C. and Thiel, M. (2012) 'Microplastics in the Marine Environment: A Review of the Methods Used for Identification and Quantification', *Environmental Science & Technology* 46: 3060–75.

Hirai, H., Takada, H., Ogata, Y., Yamashita, R., Mizukawa, K. *et al.* (2011) 'Organic Micropollutants in Marine Plastics Debris from the Open Ocean and Remote and Urban Beaches', *Marine Pollution Bulletin* 62: 1683–92.

Hopewell, J., Dvorak, R. and Kosior, E. (2009) 'Plastics Recycling: Challenges and Opportunities', *Philosophical Transactions of the Royal Society B* 364(1526): 2115–26.

Kershaw, P., Katsuhiko, S., Lee, S., Leemseth, J. and Woodring, D. (2011) 'Plastic Debris in the Ocean', in *UNEP Year Book: Emerging Issues in our Environment*, Nairobi: UNEP.

Koch, H.M. and Calafat, A.M. (2009) 'Human Body Burdens of Chemicals used in Plastic Manufacture', *Philosophical Transactions of the Royal Society B* 364: 2063–78.

Kommunenes Internasjonale Miljøorganisasjon (KIMO) (2008) *Fishing for Litter Scotland: Final Report 2005–2008*, ed. K.I. Miljøorganisasjon, Shetland: KIMO.

Laist, D.W. (1997) 'Impacts of Marine Debris: Entanglement of Marine Life in Marine Debris including a Comprehensive List of Species with Entanglement and Ingestion Records', in J.M. Coe and B.D. Rogers (eds) *Marine Debris: Sources, Impacts and Solutions*, Berlin: Springer.

Lang, I.A., Galloway, T.S., Scarlett, A., Henley, W.E., Depledge, M. *et al.* (2008) 'Association of Urinary Bisphenol-A Concentration with Medical Disorders and

Laboratory Abnormalities in Adults', *Jama: Journal of the American Medical Association* 300: 1303–10.

Law, K.L., Moret-Ferguson, S., Maximenko, N.A., Proskurowski, G., Peacock, *et al.* (2010) 'Plastic Accumulation in the North Atlantic Subtropical Gyre', *Science* 329: 1185–88.

Lithner, D., Larsson, A. and Dave, G. (2011) 'Environmental and Health Hazard Ranking and Assessment of Plastic Polymers based on Chemical Composition', *Science of the Total Environment* 409: 3309–24.

Macfadyen, G., Huntington, T. and Cappell, R. (2009) 'Abandoned, Lost or Otherwise Discarded Fishing Gear', in *UNEP Regional Seas Reports and Studies*, Rome: UNEP/FAO.

Marine Debris Solutions (n.d.) 'Plastic Doesn't Belong in our Oceans, it Belongs in Recycling Bins', marinedebrissolutions.com (accessed 20 August 2012).

Martins, J. and Sobral, P. (2011) 'Plastic Marine Debris on the Portuguese Coastline: A Matter of Size?' *Marine Pollution Bulletin* 62: 2649–53.

Meeker, J.D., Sathyanarayana, S. and Swan, S.H. (2009) 'Phthalates and Other Additives in Plastics: Human Exposure and Associated Health Outcomes', *Philosophical Transactions of the Royal Society B* 364(1526): 2097–13.

Moore, C.J. (2008) 'Synthetic Polymers in the Marine Environment: A Rapidly Increasing, Long-term Threat', *Environmental Research* 108(2): 131–39.

Moore, C.J., Moore, S.L., Leecaster, M.K. and Weisberg, S.B. (2001) 'A Comparison of Plastic and Plankton in the North Pacific Central Gyre', *Marine Pollution Bulletin* 42: 1297–300.

Mouat, T., Lopez-Lozano, R. and Bateson, H. (2010) *Economic Impacts of Marine Litter*, Shetland: KIMO.

Myers, J.P., vom Saal, F.S., Akingbemi, B.T., Arizono, K., Belcher, S., *et al.* (2009) 'Why Public Health Agencies cannot Depend on Good Laboratory Practices as a Criterion for Selecting Data: The Case of Bisphenol-A', *Environmental Health Perspectives* 117: 309–15.

Ng, K.L. and Obbard, J.P. (2006) 'Prevalence of Microplastics in Singapore's Coastal Marine Environment', *Marine Pollution Bulletin* 52: 761–67.

Oehlmann, J., Schulte-Oehlmann, U., Kloas, W., Jagnytsch, O., Lutz, I., *et al.* (2009) 'A Critical Analysis of the Biological Impacts of Plasticizers on Wildlife', *Philosophical Transactions of the Royal Society B* 364(1526): 2047–62.

Oigman-Pszczol, S.S. and Creed, J.C. (2007) 'Quantification and Classification of Marine Litter on Beaches Along Armacao dos Buzios, Rio de Janeiro, Brazil', *Journal of Coastal Research* 23: 421–28.

OSPAR (2007) *OSPAR Pilot Project on Monitoring Marine Beach Litter: Monitoring of Marine Litter on Beaches in the OSPAR Region*, London: OSPAR Commission.

PlasticsEurope (2008) *The Compelling Facts About Plastics 2007: An Analysis of Plastics Production, Demand and Recovery in Europe*, Brussels: PlasticsEurope.

——(2010) 'PlasticsEurope's Views on the Marine Litter Challenge', April, 3, www.plasticseurope.org/documents/document/20101005110.

——(2011) *Plastics: The Facts 2011 – An Analysis of European Plastics Production, Demand and Recovery for 2010*, Brussels: PlasticsEurope.

Rios, L.M., Jones, P.R., Moore, C. and Narayan, U.V. (2010) 'Quantitation of Persistent Organic Pollutants Adsorbed on Plastic Debris from the Northern Pacific Gyre's "Eastern Garbage Patch"', *Journal of Environmental Monitoring* 12: 2226–36.

Roy, P.K., Hakkarainen, M., Varma, I.K. and Albertsson, A. (2011) 'Degradable Polyethylene: Fantasy or Reality?' *Environmental Science and Technology* 45(10): 4217–27.

Ryan, P.G., Moore, C.J., van Franeker, J.A. and Moloney, C.L. (2009) 'Monitoring the Abundance of Plastic Debris in the Marine Environment', *Philosophical Transactions of the Royal Society B* 364(1526): 1999–2012.

Sathyanarayana, S., Karr, C.J., Lozano, P., Brown, E., Calafat, A.M. *et al.* (2008) 'Baby Care Products: Possible Sources of Infant Phthalate Exposure', *Pediatrics* 121: E260–68.

SCBD–STAP–GEF (Secretariat of the Convention on Biological Diversity and the Scientific and Technical Advisory Panel – GEF) (2012) Impacts of Marine Debris on Biodiversity: Current Status and Potential Solutions, Montreal, Technical Series No. 67, 61 pages.

Song, J.H., Murphy, R.J., Narayan, R. and Davies, G.B.H. (2009) 'Biodegradable and Compostable Alternatives to Conventional Plastics', *Philosophical Transactions of the Royal Society B* 364(1526): 2127–40.

Sutherland, W.J., Clout, M., Cote, I.M., Daszak, P., Depledge, M.H., *et al.* (2010) 'A Horizon Scan of Global Conservation Issues for 2010', *Trends in Ecology & Evolution* 25: 1–7.

Swan, S.H. (2008) 'Environmental Phthalate Exposure in Relation to Reproductive Outcomes and Other Health Endpoints in Humans', *Environmental Research* 108: 177–84.

Talsness, C.E., Andrade, A.J.M., Kuriyama, S.N., Taylor, J.A. and vom Saal, F.S. (2009) 'Components of Plastic: Experimental Studies in Animals and Relevance for Human Health', *Philosophical Transactions of the Royal Society B* 364(1526): 2079–96.

Teuten, E.L., Rowland, S.J., Galloway, T.S. and Thompson, R.C. (2007) 'Potential for Plastics to Transport Hydrophobic Contaminants', *Environmental Science and Technology* 41: 7759–64.

Teuten, E.L., Saquing, J.M., Knappe, D.R.U., Barlaz, M.A., Jonsson, S., *et al.* (2009) 'Transport and Release of Chemicals from Plastics to the Environment and to Wildlife', *Philosophical Transactions of the Royal Society B* 364(1526): 2027–45.

Thompson, R., Moore, C., Andrady, A., Gregory, M., Takada, H. and Weisberg, S. (2005) 'New Directions in Plastic Debris', *Science* 310: 1117.

Thompson, R.C., Moore, C., vom Saal, F.S. and Swan, S.H. (2009a) 'Plastics, the Environment and Human Health', *Philosophical Transactions of the Royal Society B* 364(1526): 1969–2166.

——(2009b) 'Our Plastic Age', *Philosophical Transactions of the Royal Society B* 364 (1526): 1973–76.

——(2009c) 'Plastics, the Environment and Human Health: Current Consensus and Future Trends', *Philosophical Transactions of the Royal Society B* 364(1526): 2153–66.

Thompson, R.C., Olsen, Y., Mitchell, R.P., Davis, A., Rowland, S.J., *et al.* (2004) 'Lost at Sea: Where is all the Plastic?' *Science* 304: 838.

United Nations Environment Programme (UNEP) (2005) *Marine Litter: An Analytical Overview*, Nairobi: UNEP.

——(2009) *Marine Litter: A Global Challenge*, Nairobi: UNEP.

US Commission on Ocean Policy (2004) *An Ocean Blueprint for the 21st Century: Final Report*, Washington, DC: US Commission on Ocean Policy.

van Franeker, J.A., Blaize, C., Danielsen, J., Fairclough, K., Gollan, J., *et al.* (2011) 'Monitoring Plastic Ingestion by the Northern Fulmar Fulmarus Glacialis in the North Sea', *Environmental Pollution* 159: 2609–15.

WRAP (2006) *Environmental Benefits of Recycling: An International Review of Life Cycle Comparisons for Key Materials in the UK Recycling Sector*, Banbury: WRAP.

——(2007) *Consumer Attitudes to Biopolymers*, Banbury: WRAP.

——(2008) *The Carbon Impact of Bottling Australian Wine in the UK: PET and Glass Bottles*, Banbury: WRAP.

——(2009) *Biopolymer Packaging in the UK Grocery Market*, Banbury: WRAP.

Yamashita, R. and Tanimura, A. (2007) 'Floating Plastic in the Kuroshio Current Area, Western North Pacific Ocean', *Marine Pollution Bulletin* 54: 485–88.

Yarsley, V.E. and Couzens, E.G. (1945) *Plastics*, Harmondsworth: Penguin.

Zarfl, C., Fleet, D., Fries, E., Galgani, F., Gerdts, G., *et al.* (2011) 'Microplastics in Oceans', *Marine Pollution Bulletin* 62: 1589–91.

Zarfl, C. and Matthies, M. (2010) 'Are Marine Plastic Particles Transport Vectors for Organic Pollutants to the Arctic?' *Marine Pollution Bulletin* 60: 1810–14.

Zubris, K.A.V. and Richards, B.K. (2005) 'Synthetic Fibers as an Indicator of Land Application of Sludge', *Environmental Pollution* 138: 201–11.

Part IV

New articulations

10 Where does this stuff come from?

Oil, plastic and the distribution of violence

James Marriott and Mika Minio-Paluello

Preamble

This chapter originally was presented as a paper to approximately 75 attendees at the Accumulation conference at Goldsmiths, University of London. The events took place in a lecture theatre on the Goldsmiths campus in New Cross, within the borough of Lewisham in South London. The specific location resonates in the story that this chapter narrates.

In the intermission before the formal paper, the presenter, James Marriott, carefully gave out Askey's cornet wafer cones filled with either Chocolate Inspiration or Light Vanilla Carte D'Or ice cream to each member of the audience. Once the ice cream had been consumed, the presentation began – with the two, now empty, ice cream tubs holding a prominent position on stage.

The chapter is structured in such a way as to retain as much as possible of the immediacy of the original presentation in which the narrative was foregrounded. It should be noted that much that is presented here is a compressed version of the fuller analysis presented in our recent book, *The Oil Road* (Marriott and Minio-Paluello 2012).

Introduction

Ice cream – which most of us clearly enjoy – epitomizes luxury food. It meets desires, it is not about satisfying a need for sustenance. At the same time, ice cream is a 'high-energy food' of a particular type, as it depends on a constant chain of refrigeration from cow to consumer. That refrigeration at the dairy, at the factory, in the delivery truck, in the supermarket and in the home consumes a substantial amount of electricity, the vast majority of which is gas or coal powered.

The make of ice cream that you have just eaten, Carte D'Or, is a standard brand. It is manufactured by the British-Dutch corporation Unilever, the world's second largest food company,[1] listed on the London Stock Exchange and headquartered in London. Unilever also produces 'ethical' ice cream under the Ben & Jerry's brand; it is 'ethical' because care is taken to determine the origins of the ingredients of the ice cream, to reduce the damage caused to the environment and workers, and to communicate these practices to the consumer.

Yet ice cream cannot exist without its containers any more than it can without refrigeration. Take it out of its box and it becomes a runny, fairly inedible and unsatisfying mess within minutes. Nowadays, the plastic container has become the delivery vehicle for the satisfaction of desire; it is now a necessary ingredient in the experience. This chapter seeks to examine the 'origin geography' of this plastic box by telling the story of its travels from source; of state and corporate attempts to control the space through which it passes; and of the differentiated violence that ensues.

Platform has been tracking these issues since the mid-1990s, examining a particular energy route since 1998, effectively unpacking 'the oil road' that lies behind the ice cream carton. The long journey of packaging from oil-bearing geology to incinerator or landfill is not just an abstracted matter of raw materials and technological transformation. We also need to examine the social and ecological processes, geopolitical interactions and economic incentives that both enable and result from this journey. While there is a literature of 'where things come from' (e.g. Böge 1993; Molotch 2003), there is relatively little research that explores the specificities of the *political* circumstances of the origin of a ubiquitous object such as these ice cream containers.

The practice of considering plastic objects 'as they are', rather than as the products of a non-plastic substance, is deeply ingrained in us. When looking at a wooden table, it is easy enough to perceive and comprehend the trees from which it was constructed. Yet, when we gaze at food packaging, few of us will visualize the crude oil and gas within it. The fact that oil and gas are drivers and motivators within global politics is widely accepted, and has been explored in a plethora of books (e.g. Friedwald 1941; Yergin 1991). Nevertheless, there has been a tendency for this understood reality to 'float' apart from the 'everyday object', and for these geopolitical dynamics to be packaged up in such generalized notions as 'addiction to oil' or 'dependency on the Middle East'. The average citizen does not have ready access to the geopolitics within these plastic boxes.[2]

Hawkins and Thompson (both in this volume) demonstrate the 'destructive after-life of plastics'. By comparison, our work traces the destructive 'pre-life' of plastics, exploring the relationships and processes of which they are a part before we come to acquire them as 'plastics'. As such, we track the life of the material of plastic objects before they are formed. We follow the passage of that material from oil-bearing rocks, through drilling rigs, pipelines, terminals, depots, refineries, factories, distribution centres and shops, to homes. We examine the impacts – both ecological and social – of that passage.

East of Eden

Our narrative today – one of many that could be told – begins at the Absheron Sill oil formation in the offshore marine territory controlled by Azerbaijan, 130 km east of Baku. Formed between 3.4 and 5.3 million years ago, 5 km under the Caspian Sea, the oil is pumped up to the Central Azeri platform, one of seven massive rigs in this part of the Caspian. The platform runs day and night, 365 days a year, pausing only for accidents, such as the September 2008 blowout that shut down the rig and forced the immediate evacuation of 212 staff.

This platform, part of a web of platforms that comprises an offshore oil infrastructure the size of a small town, is not visible on Google Earth. Perceived as offering a 'complete' satellite map of the world, there are in reality many such regions that remain blank. These platforms lie within 'the Azeri–Guneshli–Chirag PSA area', the portion of Azeri marine territory delineated by the 1994 Production Sharing Agreement signed between the Azeri state and a consortium of foreign oil companies, the Azerbaijan International Operating Company (AIOC). According to the contract terms of this agreement, the 330 km^2 of this zone are effectively controlled by the AIOC for 30 years. For citizens and non-citizens alike, this is in practice a 'forbidden zone', run by the operator of AIOC – BP, the world's fifth largest private oil corporation, listed on the London Stock Exchange and headquartered in London.[3]

This is a marine zone outside the law of Azerbaijan. Naval vessels patrol these waters, but Azeri sovereignty and laws are not fully enforced, and a powerful non-state actor exercises political power beyond control of the Azeri state. It is important to understand how power is exerted over bodies in these biopolitical zones. Individual and collective rights appear to become irrelevant, as laws and legal standards are, it would seem, perpetually suspended. A combination of the PSA regime and the improvised spatial exercise of power has created a zone where BP is ostensibly able to act with impunity, outside of the legal standards set down in Azerbaijan's constitution and legislation. Pollution can take place not only without sanction, but also without BP informing even its corporate partners, as revealed in the WikiLeaks cables (n.a. 2010).

Most of the onward journey and transformation from Caspian underwater geology to plastic container is similarly hidden from view in a number of differently created and circumscribed 'forbidden zones'. Each has evolved and been crafted to suit BP's interests; this has been accomplished by prioritizing corporate demands – under the guise of 'security' – over rights frameworks, and by placing certain geographies beyond the inspection of even the most curious citizen.

From the Central Azeri oil platform, the crude is pumped through an undersea pipeline to the Sangachal oil and gas terminal, the largest such terminal in the world outside the Middle East. Before its gates stands a huge three-sided billboard bearing the image of Heydar Aliyev, the former president of Azerbaijan and father to the current President Ilham Aliyev. The project of building the offshore platforms and extracting the deep crude has relied on, and benefited from, the stability offered by the authoritarian regime of the Aliyev family, which has ruled Azerbaijan since 1993. Bayulgen (2005) describes how:

> success in attracting foreign capital has a 'reinforcing effect' on the political regime. Authoritarian leaders solidify their hold on power by allying themselves with foreign investors … In Azerbaijan, the holdover elites from the communist era acquired the right to set the rules of the game for international oil contracts, and in that way attracted the active support of foreign capital in further concentrating their power.
>
> (Bayulgen 2005: 29)

The regime routinely imprisons dissidents and uses police to break up demonstrations. For example, a protest far from, but coinciding with, the opening of the Sangachal Terminal in May 2005 was baton-charged and many people were injured (Azerweb 2005).

After Sangachal, the crude oil is pumped through the Baku–Tbilisi–Ceyhan pipeline (BTC) across the mountains and plains of Azerbaijan, Georgia and Turkey to the Eastern Mediterranean. This pipeline runs within a geographic zone defined by policy makers in Washington, Brussels and London as an 'energy corridor'. The 1,768 km-long BTC is itself buried beneath a metre of rock and soil, and is largely invisible.

Yet, although largely unmarked, nearby areas – including whole villages – are deemed special military zones, under the control of Ministry of Interior troops (Azerbaijan and Georgia) or the militarized Jandarma police force (Turkey). Increased surveillance, restrictions on access to community land and enforced censorship of local concerns has resulted. On each of Platform's seven field visits to the area around the pipeline and to Baku itself, undertaken between 1998 and 2009, we were detained or arrested at least once. Communities and ethnic groups defined as oppositional are targeted with increased repression.

These security practices are not merely improvised behaviour on the part of local troops. Plans drawn up and decisions made in Houston, Washington and London created legal frameworks that exempted the project from any legislation stricter than the oil contract itself. BP drafted international treaties called Host Government Agreements for each country through which the pipeline passes, effectively excluding this space from the local legal regimes (Hildyard and Muttitt 2006; see also Muttitt 2011). This 1,700 km-long strip of land, including the lands and villages around it, is subject to yet another exclusion.

A short passage from our recent book *The Oil Road* serves to illustrate the experience on the ground:

Figure 10.1 Villagers in Tetriskaro in Georgia meet in the construction corridor of the Baku–Tbilisi–Ceyhan oil pipeline and its sister South Caucasus Gas Pipeline, to complain about how the heavy use of dynamite to shatter rock had also shattered their walls

Source: (Photo © Platform)

> We want to verify the existence of a military checkpoint that is said to hinder the villagers (of Tsikhisdvari) from walking up the hill to the higher pastures and the hot-sulphur springs, as they have done for generations. The passage up this track is now described by BTC as 'crossing the pipeline corridor', and therefore apparently requires an ID card and checks by troops from the Ministry of Interior. Bumping along the dirt road, we soon see the concrete wall of another oil spill catchment dam. Further up the hillside, we can see the familiar pipeline markers. Absorbed in trying to read the route of BTC across the bright green of the distant pasture, we fail to notice a sandbagged wooden hut just on our right. Suddenly we are surrounded by several angry soldiers levelling their guns at us and shouting that we must turn back. Ramazi spins the car around, and we are heading fast downhill again. Clearly the pipeline corridor is militarized here, as at other places ... The government troops working to protect BP's infrastructure have control over access not just to the route itself, but also to the meadows, forests and mountains above it. This is the outcome of the work of George Goolsby's, the lawyer at Baker Botts who oversaw the writing of Host Government Agreements concerning the state's obligations on security.
>
> (Marriott and Minio-Paluello 2012: 179)

The pipeline not only had a 'chilling effect' on human rights in its first five years of operation, but was also the cause of repression during its construction (Amnesty International 2003). In many villages on the steppe and in the hills, in the mountains and forests, communities protested against the army of machines and men that descended upon their land. In Krtsanisi in Georgia, residents blocked roads and resisted the construction work until Spetznaz special forces attacked them (Marriott and Minio-Paluello 2012).

In part, the opposition arose because communities were afraid of what would happen in the event of an accident. Could the pipeline, passing through volatile areas with latent conflict, explode if attacked?[4] Could oil leak from the pipes and pollute their soil and water? The fear of contamination of springs was particularly acute in the Borjomi National Nature Reserve, where fresh mineral water is bottled in plastic PET bottles. The distinctive-tasting Borjomi water is one of Georgia's main exports, and the forested mountains that surround the town are the Caucasus equivalent of Evian or Buxton. Opposition to the pipeline reached a crescendo in autumn 2003, with the Georgian Minister for the Environment Nino Chkhobadze refusing to issue permits, stating that BP representatives were requesting the Georgian government to violate its own environmental legislation. Later she explained: 'The pressure came above all from the company, and it was pressure not only directed toward me, but also toward the president' (Khatchadourian 2003). This geopolitical pressure exerted by the US State Department, the World Bank and BP was enough to drive the pipeline through the objections raised and browbeat Chkhobadze into submission.

Figure 10.2 Nearby residents march to Burnaz beach in south-eastern Turkey in 2009, opposing further industrialization of their coastline alongside the existing oil terminals

Source: (Photo © Platform)

The US policy framework and momentum behind this particular export pipeline were created by the Clinton Administration through the national security advisor's office in the 1990s. By the time it was constructed, BTC had been defined as a key strategic asset by both the European Union and the United States. Thus, the pipeline carries a motivating value both within domestic national politics and on the international stage. The following incident illustrates this prominence. During the August 2008 war between Russia and Georgia, Russian planes dropped bombs across BP's Baku–Supsa pipeline, sister to BTC. The Georgian government immediately claimed that the BTC pipeline itself had been attacked, trying to generate greater Western military support.

Because of its strategic importance BTC has also become a spectacular target in regional liberation struggles. Since the 1990s, the Kurdistan Workers Party (PKK) has been using military means to assert Kurdish autonomy and rights, and in August 2008 it blew up the pipeline at Refahiye, creating a fireball that burnt for six days.

All quiet on the western front?

Descending through Turkey's Taurus mountains, the oil reaches the Mediterranean coast at Iskenderun Bay. The oil companies have carved out a 'restricted marine zone' around the terminal and tanker pier which is forbidden to local residents. As a result, the fisherfolk of Golovesi and Yumurtalik are losing their livelihoods. Attempts to continue fishing have been met with fines and prison sentences – the fishing boats are easily chased down by the coastguard vessels based out of BP's marine terminal.

From this terminal, tankers take oil to all corners of the globe, including Chile, Singapore and England. However, Platform's tracking of tankers and data collation has revealed that a significant number of vessels collecting Azeri crude from Ceyhan then follow a particular route to the Italian port of Trieste.[5] One such tanker, the *Dugi Otok*, plies back and forth time after time.

This constant movement of crude across the planet's oceans and along shipping lanes is largely outside our everyday conceptual world. Apart from docking at select terminals and bunkering while waiting for the oil price to rise, these shy and awkward vessels remain elusive, largely hidden to the naked eye. However, when something goes wrong, such as a hijacking or sinking, this segment of the oil road drops its invisibility cloak. The *Brae*, the *Sea Empress*, the *Amoco Cadiz*, the *Torrey Canyon*, the *Exxon Valdez*: the names of these tankers are iconic because oil intended to remain invisible was dramatically unleashed, coming to be seen as a poisonous reality.

Figure 10.3 Turkish fisherfolk set out into the Gulf of Iskenderun, dwarfed by the *Dugi Otok* super-tanker collecting crude from the BTC Terminal near Ceyhan. The *Dugi Otok* will carry its load of oil across the Mediterranean to Muggia near Trieste in Italy

Source: (Photo © Platform)

The ecological fallout of these disasters is the vivid counterpoint to the quiet, chronic environmental and health impacts that result should the oil reach its destination *safely*: it is a tacit reminder of the plastics that end up in the oceans, as described by Gabrys in Chapter 12 and Thompson in Chapter 9. On top of this, there is a nexus of other negative products in the wake of its safe arrival, not least the emission of large volumes of carbon dioxide that inevitably will arise from its use, from its burning in refineries, power stations, car engines, jet turbines and so on. *Dugi Otok* or any of its sister ships can be seen as a climatic 'bomb', delivering up to 250,000 tonnes of carbon dioxide to be released into the atmosphere.[6]

The crude is delivered from the tanker to a terminal near Trieste, stored and then pumped into the Transalpine Pipeline (TAL), passing through north-eastern Italy, across the Austrian Alps and into southern Germany. The social disruption that accompanied the construction of the BTC pipeline in the Caucasus in the early 2000s echoed the conflicts that arose around the building of this pipeline 40 years earlier, in the mid-1960s. An energy project then on the front line of the Cold War, the construction of TAL was utilized by the Italian state as a means to expropriate land from Slovenian farmers living within Italy (Marriott and Minio-Paluello 2012: 277–81).

In contrast to the examples in Turkey, Azerbaijan and Georgia, the western section of this continuous oil road is largely unmilitarized. Walking the route of the pipeline as it passes through villages in Friuli, the Tyrol or Bavaria, there is no sign of troops patrolling the route. Today there is an unequal distribution of violence along the 'energy corridor'. The plastic that packages our 'comfort food' creates highly variegated 'discomfort' on its journey to us. The apparent calm rule of law in fossil fuel-dependent social democracies at home is built upon exported 'forbidden zones', with the ever-present possibility of violence elsewhere.

Northern lights

The vision of a consumer society, built around the calm comforts of domestic life, was central to the creation of post-war Germany. It was this vision – proclaiming to reject the social experimentation of the past 30 years – that lay at the heart of the development of cities such as Ingolstadt, just north of Munich. Here, five refineries were constructed – mostly fed by the Transalpine Pipeline – alongside power stations, plastics factories, arms manufacturers and the automobile industry. Ingolstadt is the home of Audi, with its famous strapline '*Vorsprung durch Technik*'.

Among the plastics factories is the plant at Munchmunster, now owned by LyondellBasell but originally built by BP in the 1970s. LyondellBasell is today one of the world's largest chemicals companies, listed on the New York Stock Exchange and headquartered in Houston.

The Transalpine Pipeline delivers its Azeri crude to the refinery of Neustadt just east of Ingolstadt, where the oil is broken down into a number of products, including feedstock for the Munchmunster factory, to which it travels by pipeline. The factory runs day and night, like the Neustadt refinery and the Central Azeri oil platform, only disrupted by accidents such as an explosion in December 2005 in which one fireman was killed and three workers were injured.[7]

This plastics plant produces a number of resins, including high-density polyethylenes, which are used in the stiff material for Ariel bleach bottles and, indeed, coatings for industrial pipelines. Another product manufactured by LyondellBasell is the polypropylene resin Clyrell EC340R, marketed as ideal for high-impact, low temperature-resistant food containers. It is materials such as this that are ideal for moulding into Carte D'Or ice cream boxes.

Clyrell EC340R, manufactured in Munchmunster, passes from there to moulding factories. Plastic containers manufactured there are transported to the ice cream plant in Gloucester, where they are filled with Unilever's Carte D'Or Chocolate Inspiration and Light Vanilla. From there, they are trucked to retail outlets across England, including the Sainsbury's here at New Cross. These two tubs of ice cream were purchased at the Sainsbury's supermarket near the station and carried up the hill to here. However, the exact passage of Clyrell EC340R to Sainsbury's in New Cross is relatively opaque to us. We cannot be 100 per cent sure that there is Azeri oil in these specific boxes – once again, the journey has entered a 'forbidden zone', though this time it is the globalized processes of manufacture and distribution, rather than forms of exclusion and violence, that are obscured.

Conclusion: southern comfort?

Now that these tubs have served their purpose and have successfully delivered a luxury food to us, they will be placed in the bins here in the lobby of this lecture theatre, from where they will be carried by a cleaner to the wheeled rubbish containers on the street outside and picked up by a Veolia refuse truck contracted by Lewisham Council. The lorry will take them down the hill to the South East London Combined Heat & Power waste station, where they will be incinerated to generate electricity. Azeri lithosphere will float off into London air, carrying its carbon load into the atmosphere. However, unlike the water cycle, there is no simple process to return released carbon to Caspian geology.

We could imagine an alternative afterlife for our container, in which one of us might take a tub of this ice cream to a beach on the south coast, where we might leave it by one of the rubbish bins unaware the wind would pick it up and send it tumbling across the shingle. The container would be blown into the English Channel and float off to join other plastic detritus in the North Atlantic Gyre. As Thompson describes in Chapter 9, it might sink to the ocean floor and remain there for 100, 1,000, perhaps 10,000 years. We can thus recognize a remarkable lifespan: crude oil formed 3.4 million years ago in rocks under the Caspian comes to rest on the bed of the Atlantic for the next 10,000 years. Between these two stretches is a tiny window of transformation. It might take just 22 days for Azeri oil to be transported from beneath the Caspian to the Munchmunster plastics factory. Then the container could be moulded, filled, sold and discarded in the span of the following 40 days. In the space of only two months, this oil is extracted, transported, traded, transformed and transformed again before it is sold and ultimately trashed.

At each stage of plastic's 'pre-life' before consumption, one or more corporations generate a profit, driving the process onwards. During this brief moment, it participates in numerous processes that entail, at minimum, interactions with local communities, the complexities of geopolitics and the dynamics of ecological systems – each of which is marked by violence of one sort or another.

Yet the disruptive and violent effects of these processes are largely invisible. Although these social and environmental impacts are inherent within its constitution, the plastic product in its uniformity is seemingly wiped clean of all that violence and disruption. Whether this is an 'ethical' product is not factored into composition of the plastic container. Having noted this, the aim of this chapter is not to present a demand that the pre-life of plastic be 'ethicized' in some way or other. After all, as Mazar and Zhong (2010), for instance, discuss, there are major limitations attached to ethical products.[8] Moreover, in light of the histories and trajectories narrated here, there does not seem to be much potential for an honestly 'ethical oil', despite corporate-backed attempts to claim that there is (e.g. Levant 2010).

To echo the concerns over 'the tragedy of the commons', it seems to us that this ice cream box illustrates 'the tragedy of our personal desires'. Clausewitz claimed that the hardest military manoeuvre was retreat. We need to retreat from our violence towards the global climate, and indeed the trade structures that systemically involve violence to the rights of others. We need to explore those 'ridding strategies' that Fisher examines in Chapter 6 of this volume. These strategies could be employed not to rid ourselves of consumer products that no longer hold an allure, but rather to rid ourselves of the production of consumer goods from oil.

Notes

1 *FT Global 500*, June 2011: 16, www.ft.com/cms/s/2/1516dd24-29d3a-11e0-997d-00144feabdc0.html#axzz1xr86Tcqc (accessed 20 July 2012).

2 Unravelling the global conditions out of which objects emerge has been an analytical strategy within sociology. The work of Henri Lefebvre (1968, 1974) is notable in this respect.

3 *FT Global 500*, June 2011: 15, www.ft.com/cms/s/2/1516dd24-29d3a-11e0-997d-00144feabdc0.html#axzz1xr86Tcqc (accessed 20 July 2012).

4 The BTC pipeline passes within 40 km of the 'frozen conflict' of Nagorno–Karabagk (between Azerbaijan and Armenia); within 60 km of the 'frozen conflict' of South Ossetia (between Georgia and Russia); and through north-eastern Turkey, which is an area of ongoing conflict between the Turkish state and Kurdish forces.

5 Data collated by Platform from www.marinetraffic.com/ais (accessed 20 July 2012).
6 A conservatively low estimate, based on one barrel of crude being refined into various fuels that when burned emit 317 kg of carbon dioxide. See Bliss (2008).
7 See www.icis.com/Articles/2005/12/12/1027764/basell-munchmunster-pe-shut-after-blast.html (accessed 12 December 2005). Basell Munchmunster PE shut after the blast.
8 The dangers of particular forms of 'ethicalization' are also crystallized in the following statement: 'Every time a non-governmental organization attempts to motivate change by appealing to individuals' self-interested concerns for money and status, to businesses' desire to maximize profit, or to governments' felt mandate to increase economic growth, it has subtly privileged and encouraged the portion of people's value systems that stands in opposition to positive social (and ecological) attitudes and behaviours' (Tim Kasser, quoted in Darnton and Kirk 2011: 40).

References

Amnesty International (2003) *Human Rights on the Line: The Baku–Tbilisi–Ceyhan (BTC) Pipeline Project*, London: Amnesty International, www.amnesty.org.uk/uploads/documents/doc_14538.pdf (accessed 20 June 2012).

Azerweb (2005) *Report on Monitoring Carried Out by Monitoring Group of Human Rights Organizations During the Protest Rally Held by Opposition Parties on May 21, 2005*, www.azerweb.com/ngos/417/reports/609/index.pdf (accessed 20 July 2012).

Bayulgen, O. (2005) 'Foreign Investment, Oil Curse, and Democratization: A Comparison of Azerbaijan and Russia', *Business and Politics*, intersci.ss.uci.edu/wiki/eBooks/Articles/Azerbaijan%20Oil%20Curse%20Baylugen.pdf (accessed 27 June 2012).

Bliss, J. (2008) 'Carbon Dioxide Emissions Per Barrel of Crude', 20 March, numero57.net (accessed 20 July 2012).

Böge, S. (1993) *The Well-travelled Yoghurt Pot: Lessons for New Freight Transport Policies and Regional Production*, Wuppertal: Wuppertal Institute for Climate, Environment and Energy.

Darnton, A. and Kirk, M. (2011) *Finding Frames: New Ways to Engage the UK Public in Global Poverty*, London: Bond, findingframes.org/Finding%20Frames%20New%20ways%20to%20engage%20the%20UK%20public%20in%20global%20poverty%20Bond%202011.pdf (accessed 20 July 2012).

Friedwald, E.M. (1941) *Oil and the War*, London: Heinemann.

Hildyard, N. and Muttitt, G. (2006) 'Turbo-charging Investor Sovereignty: Investor Agreements and Corporate Colonialism', in Focus on Global South (ed.) *Destroy and Profit: Wars, Disasters and Corporations*, Bangkok: Focus on Global South, www.focusweb.org/pdf/Reconstruction-Dossier.pdf (accessed 22 July 2012).

Khatchadourian, R. (2003) 'The Price of Progress: Oil Execs Muscle US-backed Pipeline through Environmental Treasure', *The Village Voice*, 22 April, www.villagevoice.com/2003-4-22/news/the-price-of-progress (accessed 20 July 2012).

Lefebvre, H. (1968) *Everyday Life in the Modern World*, London: Allen Lane.

——(1974) *The Production of Space*, Oxford: Blackwell.

Levant, E. (2010) *Ethical Oil: The Case for Canada's Oil Sands*, Toronto: McClelland & Stewart.

Marriott, J. and Minio-Paluello, M. (2012) *The Oil Road: Travels from the Caspian to the City*, London: Verso.

Mazar, N. and Zhong, C.-B. (2010) 'Do Green Products Make Us Better People?' *Psychological Science* 21(4): 494–98.

Molotch, H. (2003) *Where Stuff Comes From: How Toasters, Toilets, Cars, Computers and Many Other Things Came To Be As They Are*, New York: Routledge.

Muttitt, G. (2011) *Fuel on the Fire: Oil and Politics in Occupied Iraq*, London: Bodley Head.

n.a. (2010) 'US Embassy Cables: BP Under Fire Over Handling of Gas Leak Incident', *The Guardian*, 15 December, www.guardian.co.uk/world/us-embassy-cables-documents/171633 (accessed 20 July 2012).

Yergin, D. (1991) *The Prize: The Epic Quest for Oil, Money and Power*, New York: Simon & Schuster.

11 International Pellet Watch

Studies of the magnitude and spatial variation of chemical risks associated with environmental plastics

Shige Takada

On any visit to the beach, it is inevitable that various stranded and discarded materials will be encountered. Along the high-tide line will be seaweed, branches, trash, plastic bags, cigarette butts and more. Among this beach detritus will be plastic resin pellets (see Figure 11.1). These are small granules, generally in the shape of a cylinder or a disk with a diameter of a few millimetres. These plastic particles are the industrial feedstock of plastic products. They are produced from petroleum in chemical plants and transported to manufacturing sites, where 'user plastics' are made by re-melting and moulding. During processes of both manufacture and transport, resin pellets can be spilled unintentionally into the environment. They are then washed into aquatic environments (streams and rivers) by surface run-off during rain. Some plastics (such as polyethylene and polypropylene) are lighter than water, and therefore are not deposited but actually float on water and are carried along in streams and eventually to the ocean. In addition, pellets can be spilled directly into the ocean by the accidental dropping of batches of pellets at harbours and ports during handling or shipment. Because of the increasing global production of plastics and their environmental persistence – that is, their resistance to degradation – they are being distributed widely in oceans and are washing up on beaches and on water surfaces all over the world. The result of this increase in plastics production is that pellets are now ubiquitous. Large numbers are observed on urban beaches all around the world, from Tokyo to Los Angeles to Sydney. Pellets are also appearing on the beaches of remote islands in the open ocean in locations such as the Cocos Islands, Canary Islands, St Helens and Henderson Island. This rapid and extensive spread of plastic pellets via oceans and waterways is now a major environmental problem.

One of the main concerns with plastic resin pellets is that they carry persistent organic pollutants (POPs). POPs are human-made chemicals used in a variety of anthropogenic activities, including industry, agriculture and daily life. POPs include polychlorinated biphenyls (PCBs), different sorts of organochlorine pesticides (e.g. DDTs and HCHs) and brominated flame retardants (polybrominated diphenyl ethers, or PBDEs). Because of their very slow rate of degradation, POPs are persistent in the environment. POPs are

Figure 11.1 Plastic resin pellets

hydrophobic or lipophilic. 'Hydrophobic' is a term to express the repellence of the compound to water. Because water and oil repel each other, 'hydrophobic' and 'lipophilic' are synonyms, and both terms are used to express the affinity of the compounds to oil or oily materials. Hydrophobic compounds rarely dissolve in water, but are soluble to oil and fat. Although 'lipophilic' might be more understandable, environmental chemists prefer to use 'hydrophobic'. Though POPs are present at trace concentrations in seawater, they are much more concentrated in the fatty tissue of marine organisms such as fish and crustaceans due to their hydrophobic nature. This phenomenon is known as bioconcentration. Animals at the higher levels of the food chain are particularly vulnerable to this process, and are found to have more concentrated levels of POPs – that is, the concentrations of POPs are magnified through the food chain. Another major concern with POPs is that concentrations are biomagnified and can reach critical levels, to the point where they exhibit adverse effects on the biota. Some POPs react with DNA and damage it, causing malformation and cancer. The crossbill is an example of this malformation in wildlife. Finally, some POPs disrupt the internal system (e.g. the endocrine system) of humans and wildlife, and cause disorders of the reproductive system. This impacts on the ability to reproduce and may lead to decreased population and extinction of some species of wildlife. Some POPs affect the immune system to lower the immunity of wildlife and humans, leading to greater susceptibility to infection. For example, the mass deaths of seals that occurred in the North Sea in 1988 were ascribed to high POPs levels in the seals, which lowered their immunity and made them vulnerable to viral infection, which led to their deaths. In this case, the direct cause was viral infection, but POPs acted as the basic and indirect cause through impairment of immunity. Furthermore, effects on a wider range of internal systems (e.g. the thyroid system and brain-neurological system), and their connection with abnormalities in the behaviour of humans and wildlife, have also been suggested in recent studies (Eriksson *et al.* 2006; Schuur *et al.* 1998). In addition, POPs are transported for long distances, and in many cases cross boundaries via various routes (e.g. atmospheric transport), as discussed below. Therefore, no one government acting alone can protect its citizens or the environment from POPs; international regulation is essential.

Because of the persistent bioaccumulative nature of POPs, as well as their adverse effects and long-range transport, their production and usage have been strictly regulated by international treaty. In 2001, the Stockholm Convention was adapted in response to the global problem of POPs. As of early 2013, 185 countries and the European Union (EU) were signatories and/or had ratified the convention, which requires parties to take measures to eliminate or reduce the release of POPs into the environment. However, before these regulations were implemented, POPs were extensively released, and are now widely distributed throughout the globe. In this situation, the global monitoring of POPs is necessary to identify hotspots of contamination and to target where regulatory and remediation efforts should be concentrated. Monitoring is also important to assess the regulatory or remediation efforts. The Stockholm Convention emphasizes the importance of global monitoring, particularly in the oceans.

International Pellet Watch and persistent organic pollutants

In 1998, a colleague introduced me to plastic resin pellets and asked me to analyse the pellets for organic micropollutants. She was working in a governmental institution to gather global information on synthetic chemicals and their potential effects on the ecosystem, and became aware that marine plastics and their interactions with chemicals would become a global issue. The reason for her request was my academic background in the analytical chemistry of trace organic compounds in environmental samples. Students in my laboratory analysed the resin pellets, and we detected surprising high concentrations of organic pollutants in them. We then began studying chemicals in marine plastics, and have been conducting various studies on chemicals associated with plastics, including the identification of chemicals in marine plastics, toxic chemicals in plastic food containers, the transfer of chemicals from plastics to organisms, and the release of plastic-derived chemicals to landfill leachate. This chapter introduces International Pellet Watch and related studies on environmental plastics.

International Pellet Watch (IPW) utilizes plastic resin pellets as a tool to monitor global contamination by POPs. It developed from our finding that plastic resin pellets accumulate POPs from surrounding seawater (Mato *et al.* 2001). Plastic resin pellets are produced from crude oil, and the basic chemical structure of petroleum is retained in plastic; therefore, plastic resin pellets can be considered 'solid oil'. Plastic resin pellets have a high affinity with POPs because they easily dissolve into oil. For example, concentrations of PCBs in pellets are 1 million times higher than those in the surrounding seawater (i.e. with a concentration factor of 1 million). Another way of saying this is that five pellets correspond to 100 litres of seawater. This concentration process means that plastic resin pellets are a very useful way to investigate wider problems of serious ocean pollution, not just in terms of pollution from the pellets, but also because the pellets are carriers or transporters of other pollutants. Methodologically, collecting 100 litres of seawater and transporting it to an analytical laboratory is logistically very difficult. It is much easier to pick up five samples of pellets from beaches and harbours. Because the resin pellets are distributed on world beaches, and collection and shipping of the pellets is easy, we developed International Pellet Watch in 2005 (Takada 2006).

As part of the establishment of the International Pellet Watch project, we asked citizens across the globe to collect plastic resin pellets from nearby beaches and send them to our laboratory via air mail (see Figure 11.2). No cooling or freezing is necessary – people just have to put the pellets into an

Figure 11.2 Pellets from all over the world

envelope, indicate where they have been collected and post them to us. POPs in the pellets are then analysed in our laboratory and, based on the analytical results, the global distribution of POPs is mapped. The results are then sent to the participants via email and released on the web. The advantage of Pellet Watch is the extremely low cost of sampling and shipping compared with conventional monitoring using water, sediment and biological samples. We have been able to draw a global POPs pollution map at very low cost. We have also been able to engage non-specialists in the process of sample collection, and to give them follow-up information that makes them feel involved and connected to research on oceans and their protection. Many nongovernmental organizations (NGOs) that conduct beach clean-ups have also become involved in the project and in sample collection.

We conducted basic studies on characteristics of POPs accumulation in plastic resin pellets, and ran a pilot monitoring of POPs by using pellets on Japanese beaches with the cooperation of the Japanese NGO for beach clean-up (JEAN) in 2002 and 2003 (Endo *et al.* 2005). Based on the results, I made the first proposal for the establishment of International Pellet Watch in my presentation at the '2005 Plastic Debris, River to Sea' conference, which was held in California in September 2005. When IPW first began, only a few samples of pellets arrived. A few months after the conference, we received a couple of pellet samples from a few researchers and an NGO that had attended the conference. We took several approaches to spreading the news about the IPW project. In 2006, the editor-in-chief of the *Marine Pollution Bulletin*, Professor Charles Sheppard, provided a valuable opportunity to write an article, 'Call for Pellets', in the journal. I also gave 'Call for Pellets' presentations at several international conferences in the United States, Europe, South-East Asia and Japan. A private Japanese funding body for environmental research (the Mitsui & Co. Ltd Environment Fund), which has supported our project financially since 2008, provided the resources for me to visit South America and Africa to give talks to non-scientists calling for pellets. In all, I made the call more than 20 times in different parts of the world. Some individuals and NGOs responded to the call and sent pellets to us. We also established the International Pellet Watch website at www.pelletwatch.org in collaboration with a web company (STUDIO FLEX). Many more NGOs and individuals saw the website and sent pellets to us. Some NGOs also kindly forwarded 'calls for pellets' through their international networks. Furthermore, when we fed the analytical data back to the collectors, they were reminded of the problem of marine plastics and POPs, and forwarded the information to their friends. As a result of this activity, approximately 100 individuals from 50 countries have now joined International Pellet Watch. After six years, we now have pellets from more than 200 locations in 50 countries. Participants include researchers and governmental officers working on marine plastics, NGOs for beach clean-ups, artists who utilize marine litter to make their articles, school teachers and their students, and individuals who are simply concerned about marine environments.

Based on the analytical results of the pellet samples from beaches all over the world, we have been able to draw a global pollution map of POPs (Ogata *et al.* 2009). As an example, a map of PCBs is shown in Figure 11.3. This figure highlights various aspects of the global and local problems of POPs. PCBs were detected in all pellet samples, including those from remote islands such as the Canary Islands, Saint Helena, the Cocos Islands, the island of Hawaii, the island of Oahu and Barbados (Heskett *et al.* 2012). This indicates the wide global dispersion of POPs.

One of the most worrying features of POPs is their long-distance movement. For example, some PCBs are present in the air as vapour, and are transported by the movement of air mass from industrialized zones where the PCBs were used, to remote areas. Some portions of the aerially transported PCBs are then washed away by rain and deposited onto land and sea surfaces in remote areas in cooler periods. In warmer periods, some of these deposited PCBs evaporate, and are thus atmospherically transported further. These processes of evaporation, aerial transport and redeposition occur repeatedly, and consequently PCBs are transported far from their sources and globally dispersed. This atmospheric transport is called 'global distillation', or grass-hopping; the process is active enough to distribute PCBs globally. In addition, marine plastics may contribute to the global dispersion of POPs, as discussed below. The pellets from the remote islands tell us about the complex processes of global pollution of POPs.

In Figure 11.3, we can easily recognize hot spots of PCB pollution with much higher concentrations of PCBs in the pellets. PCB concentrations are two to three orders of magnitude higher in Los Angeles, San Francisco, Boston,

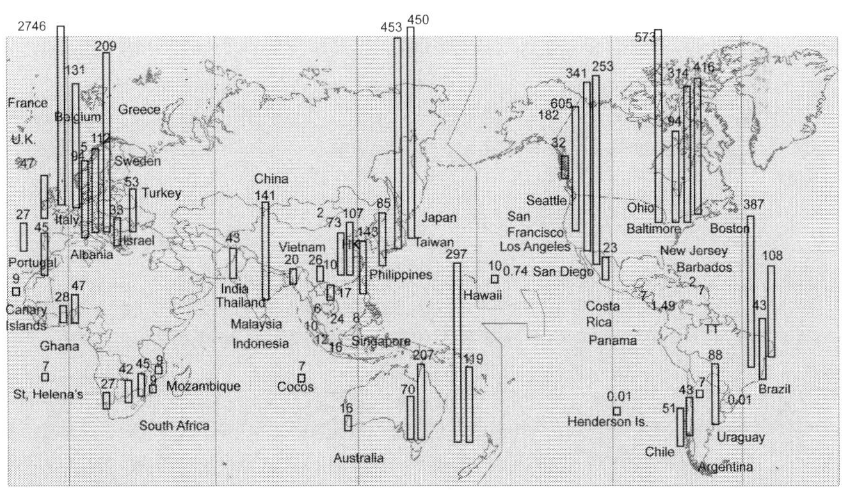

Figure 11.3 Concentration of PCBs* in beached plastic resin pellet (ng/g-pellet)
Note: * Sum of concentrations of CB#66, 101, 110, 149, 118, 105, 153, 138, 128, 187, 180, 170, 206.

Santos (Brazil), Sydney, Tokyo, Athens, the Netherlands and Normandy than in the remote areas. All of these areas are highly industrialized, and large amounts of PCBs were used in the 1960s and 1970s; some were also discharged into coastal zones and oceans. Because of their persistent and hydrophobic nature, PCBs are accumulated in the bottom sediments in the coastal zones. Usage of PCBs was banned in these countries in the 1970s, and basically there are no current inputs of PCBs to coastal waters. However, the bottom sediments containing PCBs can easily be re-suspended and remobilized by physical processes (wind, waves, currents), biological processes and anthropogenic activities such as dredging and underwater construction. Through re-suspension and remobilization, together with desorption, PCBs continue to contaminate coastal waters. These industrialized waters still suffer the legacy of PCB pollution, which is why we observe higher concentrations of PCBs in the industrialized coastal zones such as Los Angeles, Tokyo Bay, Sydney and Normandy.

On the other hand, minimal amounts of PCBs have been observed in pellets from South-East Asian countries (except for the Philippines) and from Caribbean countries. The levels of PCBs from these regions were almost the same as background levels. This is reasonable because no PCBs were used in these countries during their period of rapid economic growth in the 1980s, since PCBs were globally banned in the 1970s. However, alarming signs of new pollution are being detected in some industrially developing countries. For example, significantly higher concentrations of PCBs were detected in pellets from Ghana in comparison to global background levels. Ghana had no opportunity to utilize PCBs for industrialization because the usage of PCBs had already ceased. One plausible explanation for this anomaly is the release of PCBs from e-wastes (electronic and electric waste). Significant amounts of used electronic and electric products and waste products are imported to the industrially developing countries from industrialized countries. Some contain PCBs, which can be released into aquatic environments through the dumping and demounting or recycling of e-wastes. International Pellet Watch needs to analyse pellets from other locations in Ghana and from other countries to test this hypothesis.

Fragments of user plastics

International Pellet Watch tells us not only about the pollution status of various POPs and their global distribution, but also about risks associated with chemicals in marine plastics. Because pellets contain POPs, they act as carriers to various species of seabirds, which mistake them for food. However, plastic resin pellets are a minor component of marine plastics debris. The majority of plastics in beach debris are fragments of user plastics, as shown in Figure 11.4 and discussed elsewhere in this book. This plastic debris is also observed floating on the sea surface (Law *et al.* 2010), and can be ingested by seabirds (Yamashita *et al.* 2011). These plastic fragments also contain POPs.

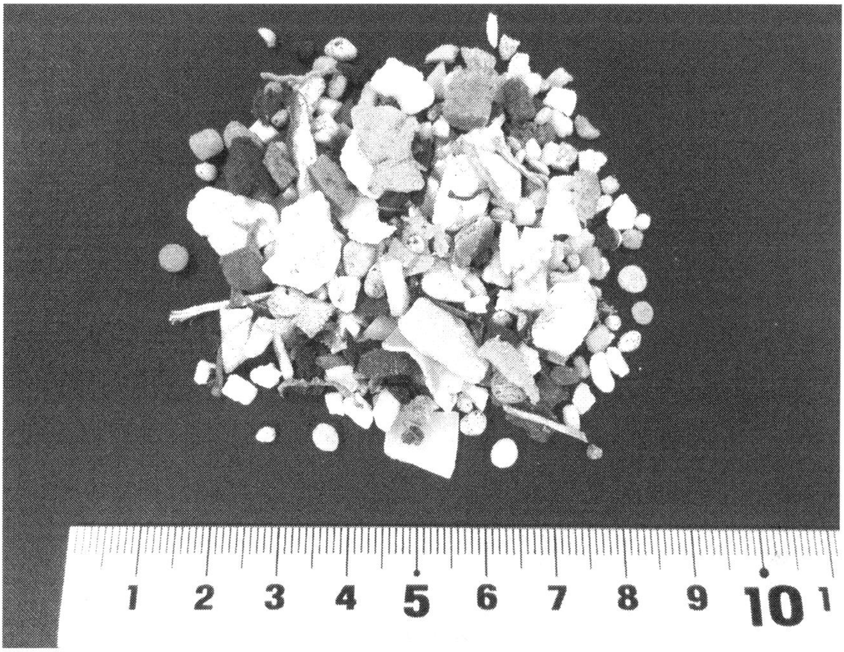

Figure 11.4 Plastic fragments from a beach on Easter Island

IPW-related study detected the same sorts of POPs in marine fragments as in plastic resin pellets (Hirai *et al.* 2011). Sorption and accumulation occur for both plastic fragments and plastic resin pellets because both consist of the same chemical components (e.g. polyethylene, polypropylene), and therefore have a similar hydrophobic nature that attracts POPs. All marine plastics – that is, plastic resin pellets and plastic fragments – contain various POPs and similar pollutants, including PCBs, organochlorine pesticides, brominated flame retardants, petroleum hydrocarbons and polycyclic aromatic hydrocarbons. In addition to seabirds, various other marine organisms ingest marine plastics. Teuten *et al.* (2009) have documented more than 180 species of animals that have ingested plastic debris, including birds, fish, turtles and marine mammals. Small plastic pieces floating on the ocean surface are mistaken for food by both fish and birds, while turtles eat suspended plastic bags, which they may mistake for jellyfish. This shows how marine plastics operate as accumulators and carriers of POPs to marine organisms.

Marine plastics: a new vehicle for transporting POPs in marine environments

This process of transporting POPs in plastic has been a key finding of the International Pellet Watch research. However, the movement of POPs by marine plastics – which are defined as plastics with a size ranging from 1 mm to 10 cm, and a density lighter than water – is a new concept. Traditionally, it was thought that POPs were carried in aquatic environments by natural particles, including soil particles, soot, phytoplankton, faecal materials and the debris of marine organisms. These particles are deposited onto bottom sediments, and it was assumed that they did not carry POPs over a long distance. Traditionally, POPs have accumulated in sediments in the vicinity of their source of production or spillage. However, the steady growth of the variety and amount of marine plastics that are non-degradable and floating has shown that they have unique characteristics as carriers of POPs. Some researchers have shown that marine plastics travel long distances (Moore *et al.* 2001; Law *et al.* 2010). For example, plastic fragments discharged from Japan were found on a beach in Midway Atoll, indicating that marine plastics can be transported over 4,000 km. Marine plastics can carry POPs for longer distances than natural particles.

In the case of this movement, the question then emerges as to whether the sorption of POPs into marine plastics is evidence of an equilibrium process and whether POPs might be desorbed from plastics during transport in the open ocean. This is true: POPs that have accumulated in marine plastics can be released from plastic during transport in cases where POPs concentrations in the ambient seawater are lower. The accumulation of POPs in plastics is not a one-way process, but rather *bidirectional*. POPs are also released from plastics into open ocean water, where POPs concentrations in seawater are lower than in the industrial coastal zone. This process is called desorption. However, it is quite a slow process in marine plastics and can take from several months to years. A recent study has confirmed the 'slow' desorption in our latest field experiments (Rochman *et al.*, 2013). One plausible mechanism behind the slow desorption is migration (or movement) of POPs in a plastic particle matrix. Sorption and desorption of POPs into plastic is not a surface phenomenon, but is controlled by the migration of POPs into the particle matrix (Karapanagioti *et al.* 2010). That is, POPs penetrate into the inside of plastic particles on sorption, whereas POPs should be moved from the inner part of a plastic particle to the surface on desorption. On sorption/desorption, POPs have to move within the plastic matrix. The speed of movement is slow, and therefore it would take time for POPs to be released from plastic particles to ambient seawater because a plastic particle has a relatively longer diameter (a few millimetres) compared with natural particles, the diameter of which is around a few micrometres. In the case of natural particles, migration of POPs into the particle matrix takes a shorter time (hours) because of the shorter diameters. As a result, desorption of POPs from marine plastics takes months to years, while desorption from natural particles takes only a few hours. The detailed mechanism of slow desorption is still under investigation; the main finding so far is that the process is very slow, and some POPs can be carried for a long distance before they are completely leached out from plastics. As a result of this issue of size and speed of desorption, we have observed some plastic pieces that contain higher concentrations of POPs than the other plastics found on remote islands. In the International Pellet Watch project, we always analyse five plastic samples from each location and have found some plastic samples with sporadically high concentrations of POPs in even the remote areas. For example, PCB concentration in a sample from a remote beach in Uruguay exhibited 24 nanograms per gram (ng/g), whereas the other four samples had much lower PCB concentrations of less than 0.07 ng/g (see Figure 11.5). In International Pellet Watch, we measure median concentration to exclude sporadic high concentrations and to get a representative pollution status for individual locations. Therefore, PCB concentration at this location in Uruguay was 0.01 ng/g. This is reasonable because it is a remote area without nearby industrial activities. However, marine organisms cannot take median doses – they take everything! Thus the plastic sample with sporadically high concentration of PCBs (i.e. 24 ng/g) poses a serious threat to marine organisms. In remote areas, wild animals are exposed to minimal amounts of POPs through natural media (water, air and the food web). Wildlife are also vulnerable to the threat of POPs through sporadic high concentrations. The existence of sporadic high concentrations of POPs is a unique but hazardous aspect of marine plastics. Some scientists indicate that the transport of POPs by marine plastics is less important than natural particles. Their model for this calculation is based on homogeneous distribution of POPs among marine plastics. However, the real world is heterogeneous and far more complex. Sporadic high concentrations of POPs in pellets have been

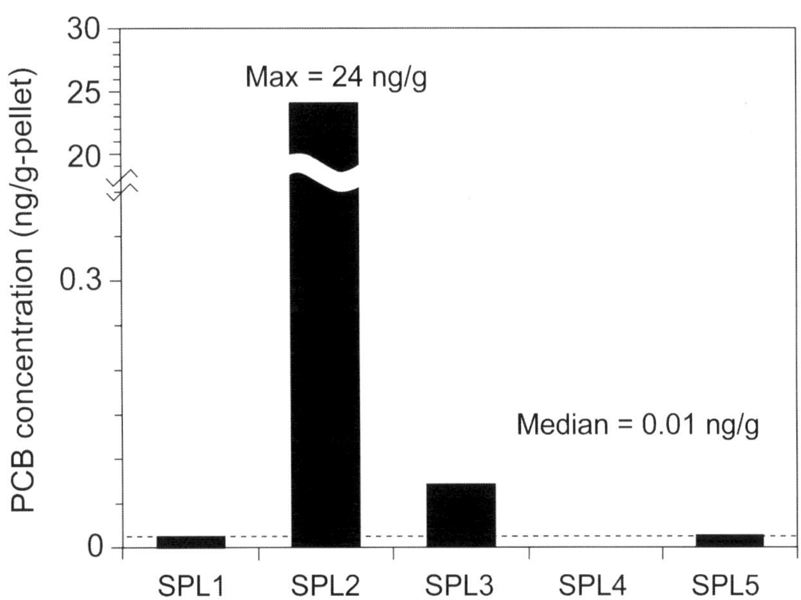

Figure 11.5 PCB concentrations in five samples from a remote beach in Uruguay

observed at many other locations beyond Uruguay. International Pellet Watch provides real-world data to assess the risk of plastic-mediated exposure of chemicals to marine organisms.

The transfer of POPs to seabird tissue

As discussed above, marine plastics can carry POPs to marine organisms. In IPW-related studies, we were also interested in finding out whether POPs were transferred to the tissue of marine organisms. We got weak but significant evidence of the plastic-mediated transfer of PCBs to seabirds. One reason why our evidence is ambiguous is because marine organisms are exposed to POPs through their natural prey, and POPs in prey normally are magnified in their concentration through the food web. That is, POPs are brought to marine organisms both through their natural prey and through marine plastics (see Figure 11.6). When we focused on the POPs with lower concentrations in natural prey and higher concentrations in marine plastics, we found evidence of a transfer of POPs from marine plastics to marine biota. We measured concentrations of a kind of PCB (CB#8; 2,4'-dichlorobiphenyl) in the fatty tissue of one kind of seabird, the Short-tailed Shearwater. These bird samples were collected by catch in the north Pacific. They were accidentally trapped in fishing nets and killed during experimental fishing. All 12 seabird samples contained plastics in their stomachs to differing degrees. Significant correlation was observed between the amounts of plastics in the birds' stomachs and concentrations of the PCB (CB#8) in the fatty tissue of the seabird (Yamashita *et al.* 2011). This shows that the seabirds that ingested more

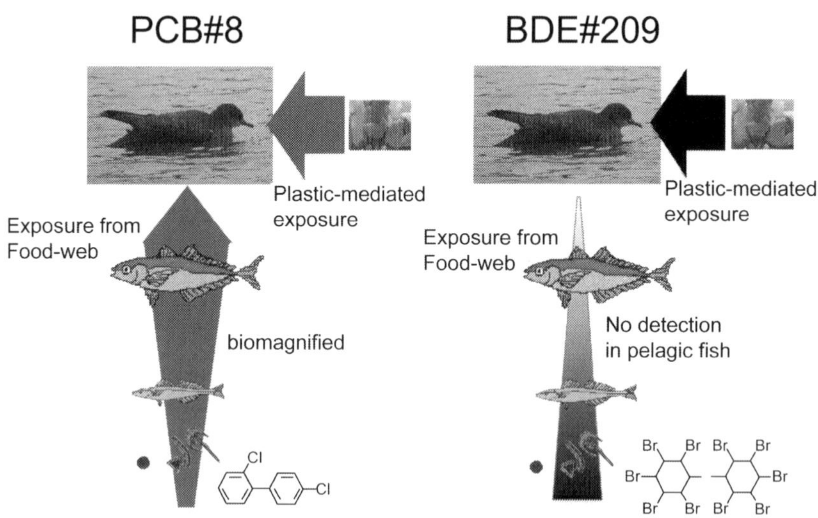

Figure 11.6 Models of chemical exposure on seabird via food-web and ingested plastics

plastics had more PCBs in their fatty tissue. It also suggests that there is a transfer of PCBs from ingested plastics. Similar evidence was obtained for a different species of seabird in a different area by Ryan, Connel and Gardner (1988). In our research, we also found other evidence of the transfer of PCBs (Teuten *et al.* 2009). We conducted a feeding experiment where another species of seabird, the Streaked Shearwater, was fed with plastic resin pellets from Tokyo Bay that were contaminated with PCBs. Concentrations of PCBs in oil excreted from the preen gland of the seabird (preen gland oil) were measured every week during the experiment. Concentrations of sorts of PCBs (lower chlorinated congeners) in preen gland oil increased through Day 7 for the plastic-feeding setting, whereas no such increase was observed for the control group. The increase was slight but significant. This is due to the fact that marine organisms are exposed to PCBs not only through plastics but also through the food web (Figure 11.6). When we focus on other compounds that are not contained in natural prey, we can get clearer evidence.

Recently, we obtained evidence that one sort of plastic additive, polybrominated diphenyl ethers (PBDEs), is transferred to the tissue of seabirds that ingest marine plastics (Tanaka *et al.* n.d., 2013). PBDEs are compounded to some plastics as flame retardants, and they have the potential to disrupt the thyroid systems of biota. PBDEs were measured in the abdominal adipose tissue of the same Short-tailed Shearwater individuals that were studied for PCBs. Ingestion of plastics by this species of seabird is frequently observed. Plastic fragments were detected in the digestive tracts of all 12 individuals examined. Our analysis of the plastic fragments from the Short-tailed Shearwater demonstrated that plastic fragments from two individuals contained a sort of PBDE – BDE#209, which is a major component of currently used flame retardants. Sporadic detection (i.e. 2 from 12 samples) reflects the nature of additives – that is, some plastic products contain the flame retardants for specific use of the products but the others do not contain them. BDE#209 was also detected sporadically from plastic fragments collected from the sea surface in open ocean (Hirai *et al.* 2011). On the other hand, BDE#209 is normally not detected in pelagic fish. Exposure of seabirds to BDE#209 through the food web (e.g. fish) is unlikely; however, BDE#209 was actually detected in the tissue (abdominal adipose) of two seabird individuals, meaning it must come from the ingested plastics. Interestingly, the two individuals bore BDE#209 in both ingested plastic and the internal tissue. These findings suggest evidence of transfer of chemicals from ingested plastics to the organisms, though more observations are needed. We also have to address the question of whether these degrees of exposure of plastic-derived chemicals (e.g. BDE#209) may cause adverse effects in the organisms (e.g. seabirds).

These are the issues with which we are currently involved. In this section, I've talked about one kind of plastic additive: flame retardants in marine organisms. However, plastic additives can also sometimes cause problems directly for humans, as discussed in the next section.

Plastic-derived nonylphenols: here, there and everywhere

To maintain the performance of plastics, various additives such as plasticizers, antioxidants, anti-static agents and flame retardants are compounded into plastic products. Some of these plastic additives are leached from plastic product waste, and may threaten humans and wildlife. An example is the release of nonylphenols, or endocrine-disrupting chemicals, from plastic tubes that are employed for scientific research. This is described in a chapter entitled 'Here, There, and Everywhere' in a well-known book on endocrine disruption, *Our Stolen Future*, by Colborn, Dumanoski and Meyers (1996).

Drs Anna Soto and Carlos Sonnenschein at the Tafuts Medical School in Boston studied the multiplication of cells. One day they incidentally encountered disruption of their experiments by nonylphenols leached from a plastic lab tube. Nonylphenols are a kind of organic chemical that consists of a benzene ring, a hydroxyl functional group and a branched alkylchain containing nine carbons. They are added to lab tubes as antioxidants to give a more stable and less breakable nature to these plastics. Drs Soto and Sonnenschein observed that the nonylphenols leaching out of the plastic tube caused rampant proliferation of breast cancer cells. Their observation showed that the chemicals leaching out of plastic mimicked oestrogen and resulted in the crazy multiplication of the breast cancer cell. This is a clear example demonstrating that additives released from plastics potentially disrupt the endocrine system of humans. Many follow-up studies have demonstrated that nonylphenols are endocrine-disrupting chemicals. In the normal endocrine system, oestrogen is secreted at appropriate time points in appropriate individuals. It binds with oestrogen receptors and activates DNA in the nucleus to produce appropriate biological responses. However, a range of human-made chemicals can bind with oestrogen receptors when they enter the cells of humans and wildlife. The association of chemicals with oestrogen receptors gives inappropriate signals at inappropriate times to induce inappropriate biological responses. This initiates abnormal behaviours in the endocrine system and can cause disorders in the reproductive system and lower the effectiveness of reproduction. This leads to diseases caused by imbalance in oestrogen, such as vaginal clear cell adenocarcinoma, as well as a decreased ability to reproduce. These phenomena are referred to as 'endocrine disruption'. Nonylphenols are typical endocrine-disrupting chemicals. Some other endocrine-disrupting chemicals, such as Bisphenol-A, are also derived from plastics and are explored below.

Based on concerns derived from this study, we analysed various objects made of plastic and found nonylphenols in various items. Even in ultra-pure water, we detected nonylphenols when we stored it in a polyethylene tank. A more serious concern, however, was the exposure of humans, and especially babies, to endocrine-disrupting nonylphenols derived from plastics. When I read the findings of Drs Soto and Sonnenschein, my twin nephew and niece were seven months old and had just started to use plastic teethers. I found that the teethers were made of silicon (a kind of polymer), and was concerned about the potential of endocrine-disrupting chemicals leaching out. In analysing the teethers, I did detect nonylphenol, so I advised my sister-in-law not to use the teethers. I also called the company and advised it not to sell the teether. It did not know that the product contained the endocrine-disrupting chemicals and said it would stop the import and sale of the product.

Another concern that we investigated in IPW-related studies was whether nonylphenols were detected in a wider range of consumer plastics. We analysed 50 plastic products available on the Japanese market, including food containers, cups and dishes. Again, high levels of nonylphenols were detected in several items, including disposable transparent plastic cups (Isobe *et al.* 2002). Other research detected nonylphenols in plastic wraps (Kawanaka *et al.* 2000; Funayama *et al.* 2001). These researchers detected nonylphenols in rice balls that were wrapped with the plastic wrap, indicating the transfer of nonylphenols to the rice ball (Funayama *et al.* 2001). Nonylphenol was also detected in the plastic containers of snacks and sweets, and transfer to ice cream was confirmed. As Colborn, Dumanoski and Meyers (1996) argue, plastic-derived endocrine-disrupting chemicals are literally here, there and everywhere. Based on these findings, and public concern about endocrine-disrupting chemicals, some governmental sectors in Japan recommended that plastic industries not use nonylphenol-based additives in their consumer plastics. This official recommendation appears to have been effective, and only trace amounts of nonylphenol were detected in the plastic products made in Japan in 2010.

Despite this regulation of and reduction in use in Japan, the problem is international. Because of acceleration of trading, many low-priced plastic products are imported to Japan from China and South-East Asian countries. We surveyed 100 imported and domestic (i.e. Japanese) plastic food containers, bags and related items in 2010. Nonylphenols were detected in some plastic products from China and Thailand, though there was almost no detection for Japanese products. It is our feeling that international regulations or guidelines should be established regarding the addition of nonylphenols and related compounds to all food and drink containers.

One of the most popular plastic containers for water and other beverages is polyethylene terephthalate, or the PET bottle. PET does not contain nonylphenols. However, we were suspicious about the caps of these plastic bottles. They are normally made of polyethylene, though some are made of polypropylene. We analysed 93 caps of mineral water bottles purchased in various countries, including Japan, South Korea, China, Vietnam, Indonesia, Malaysia, Germany, the United Arab Emirates (UAE), South Africa, Kenya and Ghana. Nonylphenols were detected in 44 of the 93 caps. Again, no or only trace amounts of nonylphenols were detected in Japanese caps, whereas higher and significant concentrations of nonylphenols were detected in caps from China, Indonesia, Korea, Malaysia, the UAE, Germany, France, USA and Kenya (see Figure 11.7). Concentrations were up to 200 ng/g. Though we

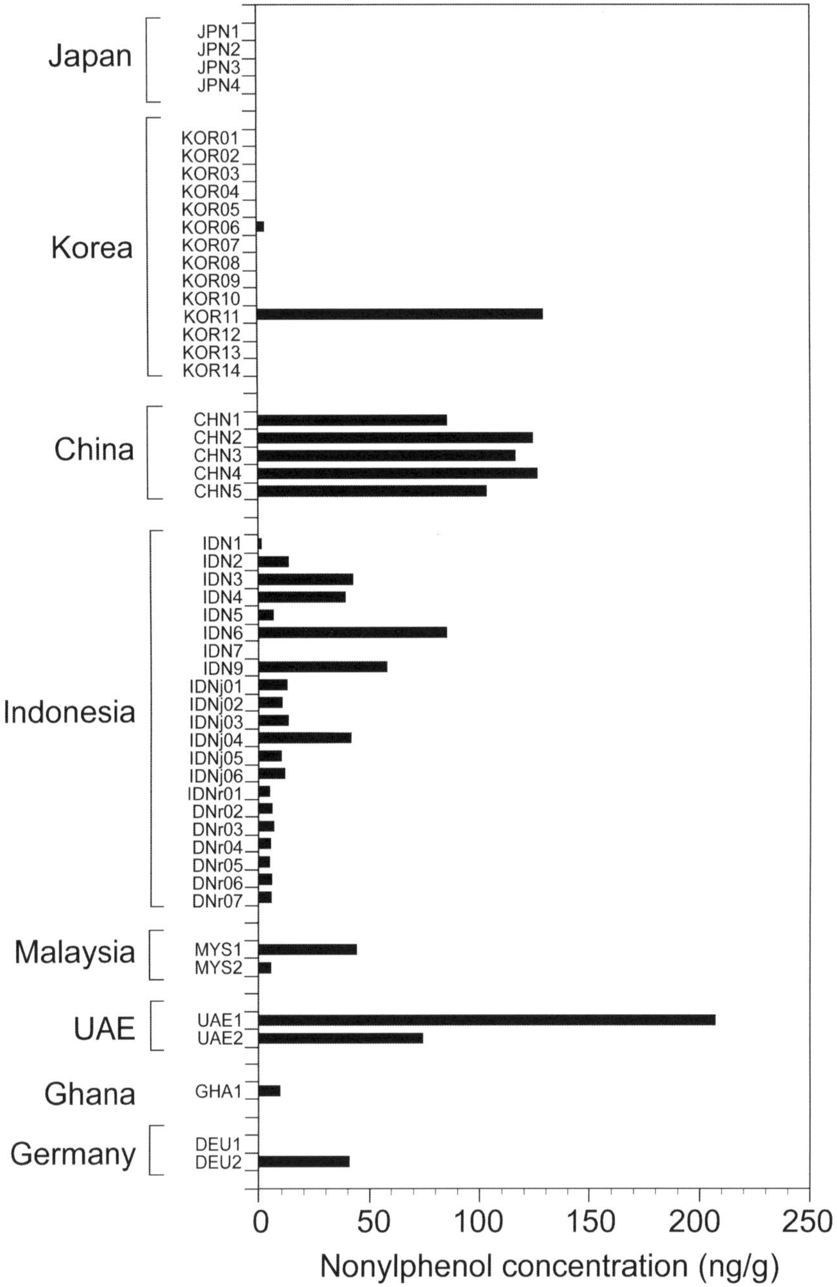

Figure 11.7 World Cap project: nonylphenol concentrations in plastic caps of bottles of mineral water purchased in various countries

have not examined the leaching of nonylphenol from the caps to the contents, several reports have detected nonylphenols in bottled water (Toyooka and Oshige 2000; Amiridou and Voutsa 2011). Bottled water is becoming increasingly important as a source of drinking water in those countries where a safe water supply system is not established. However, while bottled water might be microbiologically safe, it is not always chemically safe.

Mobile sources of endocrine-disrupting chemicals in marine environments

Plastic bottle caps containing nonylphenols are a potential threat not only to humans but also to wildlife. Plastic caps are some of the most frequently detected debris on beaches. They are as ubiquitous as resin pellets. International Pellet Watch analysed 58 plastic caps collected from several Japanese beaches (see Figure 11.8a). Among seven caps collected on a beach in Tokyo Bay, only one sample contained nonylphenols. This is reasonable because Tokyo Bay is a semi-enclosed bay and it is mostly Japanese plastic caps, which contain trace amounts of nonylphenol as indicated above, that are found on the beaches. On the other hand, nonylphenols were more frequently detected in plastic caps collected on beaches in the western islands of Japan, which are more exposed to ocean currents from China (see Figure 11.8b). A total of 28 caps among 45 samples on the western islands contained

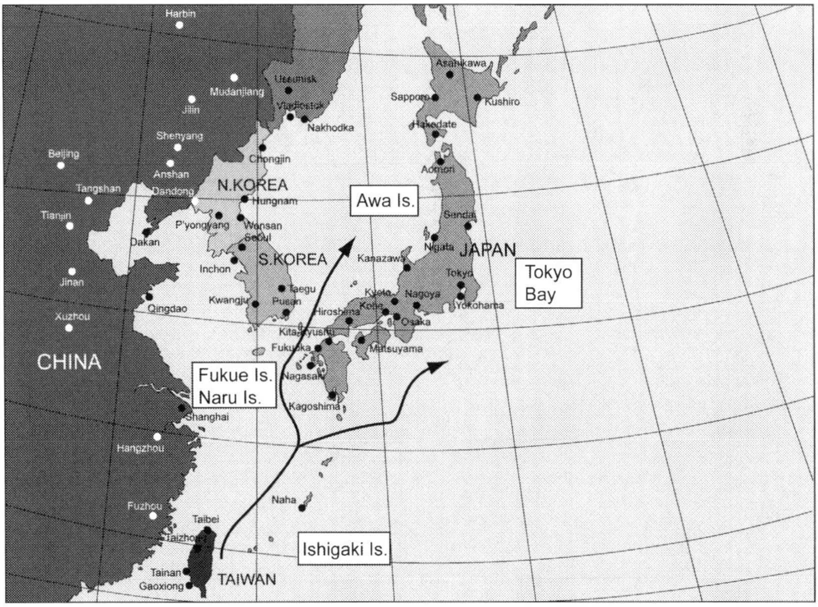

Figure 11.8a Sampling locations of beached plastic caps

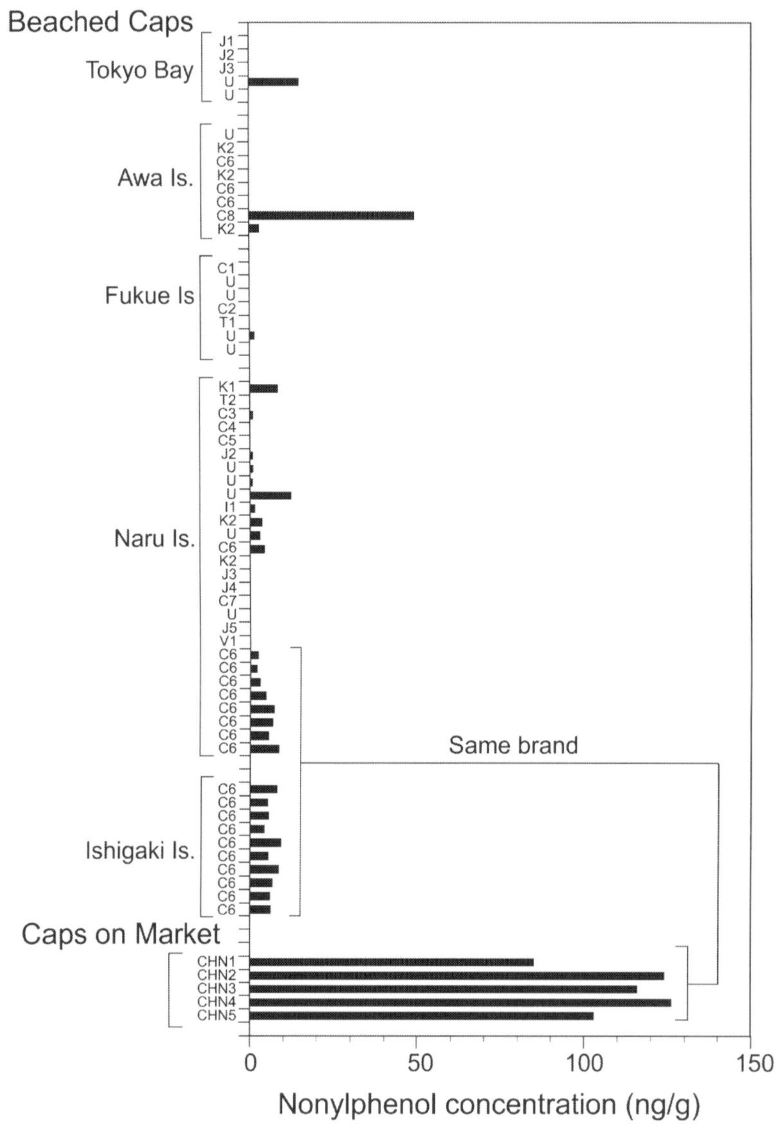

Figure 11.8b Nonylphenol concentrations in plastic caps from markets and beaches

nonylphenols. Because some species of biota, such as the Albatross, ingest this sort of plastic debris, these plastic caps could be a source of internal exposure to endocrine-disrupting chemicals. In addition, these caps are often fragmented by environmental weathering into smaller pieces, which can be ingested by a wider variety of marine species. The western islands of Japan accumulate a lot of Chinese plastic debris because they are located downstream of the Kuroshio ocean current. We observed larger quantities of Chinese plastic debris, including plastic caps from bottled water, on the beaches in the western islands. How did we distinguish Chinese debris from Japanese debris? Chinese characters imprinted on the plastic caps made the caps originating from China easy to recognize. We also used specific imprinted characters on the caps from specific brands in identification. Through imprinted characters, products on the market and debris on the beaches were interconnected. Figure 11.8b compares commercial products on the market and beached plastic caps regarding the same brand of plastic caps. Though significant concentrations of nonylphenols were detected in all the beached plastic caps, their concentrations were two orders of magnitude lower than those in the commercial products purchased on the market. This can be explained by the fact that the nonylphenols were leached out into seawater during their journey across the East China Sea. In other words, plastic caps release nonylphenols to seawater while floating. Plastic caps, like resin pellets and other plastic debris, act as mobile sources of the endocrine-disrupting chemicals to marine environments.

Though plastic caps for water bottles are most frequently observed on beaches, they are still just a minor amount of all beached plastic debris. The majority of fragments are from a wide range of user plastics. These other plastic fragments collected from marine environments, including the open ocean and remote islands, also contain nonylphenols (Hirai *et al.* 2011). In the beached plastic fragments, other additives – such as Bisphenol-A and polybrominated diphenyl ethers (PBDEs) – were detected. These also have endocrine-disrupting effects. Marine plastics are mobile sources of a wide spectrum of endocrine-disrupting chemicals.

Scavenging plastic-derived chemicals in landfills: problems with plastics in terrestrial environments

So far, we have focused on plastics in marine environments and the issues surrounding them. However, problems associated with plastics occur not only in marine environments but also on land. Considerable and growing amounts of plastics are disposed of in municipal landfills. Rainwater washes over the dumped plastics, and certain additives and monomers are released from them, thus adding to the chemicals that are found in landfill leachate (see Figure 11.9). A key sign of economic growth and industrialization is larger amounts of plastics in society, with consequent increases in the amount of plastics waste.

To investigate the effect of industrialization on the presence of endocrine-disrupting chemicals in landfill leachates, we measured plastic-derived chemicals in leachates from tropical Asian countries, including Malaysia, Thailand, the Philippines, Vietnam, Cambodia, Laos, India and Japan at different stages

Figure 11.9 Leaching plastic-derived chemicals in landfill

of economic growth (Teuten *et al.* 2009). Though high concentrations of nonylphenol were detected, Bisphenol-A showed the highest concentrations in the leachate from tropical Asian countries, ranging from 0.18 to 4,300 micrograms per litre (μg/L) (see Figure 11.10). Bisphenol-A concentrations were one to five orders of magnitude higher than those in sewage effluents. Bisphenol-A in leachate could be derived from unreacted monomers in disposed polymers (polycarbonates and epoxy-resins), and degradation of the polymers and additives. In many landfill sites in industrialized countries, treatment facilities are installed and the environmental burden of these endocrine-disrupting chemicals is reduced. High removal efficiency (95–99.7 per cent) of BPA has been reported. However, because of high concentrations of Bisphenol-A in raw leachates, even treated leachates showed higher BPA concentrations (0.11–30 μg/L; see Asakura *et al.* 2004; Wintgens *et al.* 2003) than the no-effect concentration of Bisphenol-A for endocrine disruption in freshwater organisms. For example, Bisphenol-A at 0.008 μg/L was reported to induce malformations in the female organs of the freshwater snail, *Marisa cornuarietis* (Schulte-Oehlmann *et al.* 2001). Bisphenol-A is more problematic in developing Asian countries' landfill sites because they have no or poorly functioning leachate treatment facilities. Consequently, high concentrations of Bisphenol-A are discharged into the surrounding environment, such as rivers and groundwater. For example, Bisphenol-A concentrations in water samples

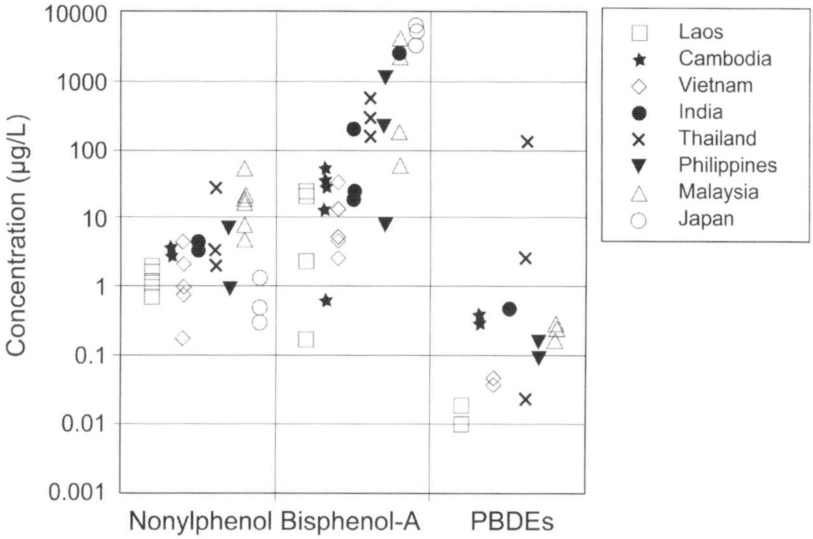

Figure 11.10 Concentrations of plastic additives: nonylphenols, Bisphenol-A and polybrominated diphenyl ethers (PBDEs) in leachate from landfills of Asian countries

from a Malaysian pond, into which the leachate from the dump flowed, were an order of magnitude higher (i.e. ~11 μg/L) than in the upstream inflowing river (0.45 μg/L) (Teuten *et al.* 2009). This clearly demonstrates that plastic waste-derived chemicals significantly increase the concentrations of endocrine-disrupting chemicals in the environment.

Among the countries investigated, the more industrialized countries – for example, Malaysia and Thailand – had higher Bisphenol-A concentrations in landfill leachate than less industrialized countries, such as Laos and Cambodia. Bisphenol-A concentrations in the leachate showed a significant positive correlation with the per capita gross domestic products (GDP) of tropical Asian countries ($r^2 = 0.66$, $n = 26$, $P < 0.0001$). The most probable reason is that more industrialized countries use larger quantities of plastics, resulting in the generation of more plastic waste. This suggests that economic growth in developing countries may increase the environmental prevalence of endocrine-disrupting chemicals unless the leachate is collected and properly treated.

Another kind of plastic additive, brominated flame retardants (i.e. PBDEs), was detected in high concentrations of up to 133 μg/L in the landfill leachate from tropical Asian countries (Kwan *et al.* n.d., forthcoming). Again, this is evidence that high concentrations of PBDEs are found in the more industrialized and populated tropical Asian countries like Thailand, the Philippines and Malaysia. These concentrations were higher than those normally observed in water and sediments from rivers, canals and sewage, indicating that leachates are a serious potential source of PBDE contamination in the environment. Our detailed examination of the composition of PBDEs demonstrated that less toxic types of PBDEs are transformed to more toxic types of PBDEs in anaerobic conditions in the landfills. The garbage dumping sites act as an amplifier of the toxicity of plastic additives through anaerobic transformation, and as emitters of the toxic chemicals to the environment.

Conclusions

This chapter has demonstrated that plastics can act as exposure sources of potentially toxic chemicals to humans, environments and wildlife through various routes. The research conducted for the International Pellet Watch study is ongoing. The key findings so far are that marine plastics (i.e. resin pellets and fragments) act as transporters of POPs and have fundamentally been involved in making these serious pollutants mobile and dispersing them far and wide. We have also suggested the ways in which various plastic additives such as nonylphenols, which are contained in some plastic caps, can be released into bottled water. More seriously, plastic caps act as mobile sources of the endocrine-disrupting chemicals to enter seawater through desorption. Plastic debris can also function as an internal exposure source for a spectrum of chemicals such as Bisphenol-A and brominated flame retardants when these plastics are ingested by marine organisms. When plastics are dumped in landfill sites, these additives are released into leachate, which contaminates groundwater, rivers and coastal waters. We also found some evidence of the transfer and accumulation of the plastic-derived chemicals in the tissues of organisms that ingest the plastics, though more concrete evidence is necessary. We hope to investigate and assess the magnitude of plastic-mediated exposure and its adverse effects on marine ecosystems in future studies. Plastic is necessary in modern society; however, International Pellet Watch and the related studies show that its afterlife is complex and often very hazardous to both humans and the environment. To reduce the burden of endocrine-disrupting chemicals, more regulations and international coordination are needed. The Stockholm Convention was a major breakthrough, and had a very positive effect on stopping the production of POPs. We need more conventions such as this, as well as a significant reduction in the unnecessary use of plastic in order to protect the world from the toxicological dangers of plastic debris.

To reduce the environmental burden of plastics and associated chemicals, growing public awareness of the problems associated with plastics is an important key. Awareness of the problems changes behaviour of individuals to reduce the usage of single-use plastics such as plastic bags for shopping and plastic bottles for drinking water. Increased public awareness also pushes national and international policies forward. International Pellet Watch could help to increase understanding of the problems of chemicals associated with plastics. The global pollution maps clearly demonstrate that plastic debris carries toxic chemicals. Furthermore, close examination of the data of

International Pellet Watch (see Figure 11.5) shows that marine plastics can carry pollutants even to remote areas. Some NGOs are utilizing the information on the IPW website to increase public awareness of the plastic issue. Most recently, as a part of such activity, we organized an international symposium on marine plastics pollution in Tokyo. International Pellet Watch has given a message of no single-use plastic to the participants.

Reflecting IPW in policies would be somehow indirect. Over the past few years, some governmental (e.g. the National Oceanic and Atmospheric Administration, or NOAA, in the United States) and international agencies, such as the Joint Group of Experts on the Scientific Aspects of Marine Environmental Protection (GESAMP), have held workshops on marine plastics to discuss the magnitude of the problems of marine plastics and the possible solutions. Results from International Pellet Watch have been referred to and discussed in the workshops and published in their official reports (e.g. GESAMP 2010). In addition, the United Nations Environment Programme (UNEP) features some key topics and publishes them in the UNEP *Yearbook* each year. In the 2011 *Yearbook*, UNEP featured marine plastics and IPW, and a global pollution map. These agencies are still discussing how they will incorporate the problems of marine plastics into their policies. I believe that the results of IPW will be properly reflected in future policies.

Acknowledgements

The author expresses sincere appreciation to all the participants of International Pellet Watch. The author thanks Dr Rei Yamashita, Mr Masaki Yuyama, Ms Bee Geok Yeo, Ms Maki Ito, Ms Yuko Ogata, Mr Satoru Iwasa, Mr Kosuke Tanaka, Mr Hisashi Hirai, Ms Chikako Aoki, Ms. Kaoruko Ichikawa, Dr Satoshi Endo and Ms Charita Kwan for providing unpublished data; Dr Kaoruko Mizukawa for providing toxicological information on POPs; Drs Yutaka Watanuki and Atsuhiko Isobe for providing precious samples and comments; Dr Takao Katase for providing instrument for nonylphenol analyses; the Mitsui & Co. Ltd Environment Fund and Thermo Fisher Scientific for their continuous financial support. The author greatly appreciates Dr Gay Hawkins's work in editing the draft.

References

Amiridou, D. and Voutsa, D. (2011) 'Alkylphenols and Phthalates in Bottled Waters', *Journal of Hazardous Materials* 185: 281–86.

Asakura, H., Matsuto, T. and Tanaka, N. (2004) 'Behavior of Endocrine-disrupting Chemicals in Leachate from MSW Landfill Sites in Japan', *Waste Management* 24: 613–22.

Colborn, T., Dumanoski, D. and Meyers, J.P. (1996) *Our Stolen Future: Are We Threatening Our Fertility, Intelligence, and Survival? A Scientific Detective Story*, New York: E.P. Dutton.

Endo, S., Takizawa, R., Okuda, K., Takada, H., Chiba, K., *et al.* (2005) 'Concentration of Polychlorinated Biphenyls (PCBs) in Beached Resin Pellets: Variability Among Individual Particles and Regional Differences', *Marine Pollution Bulletin* 50: 1103–14.

Eriksson, P., Fischer, C. and Fredriksson, A. (2006) 'Polybrominated Diphenyl Ethers, a Group of Brominated Flame Retardants, Can Interact with Polychlorinated Biphenyls in Enhancing Developmental Neurobehavioral Defects', *Toxicological Sciences* 94: 302–9.

Funayama, K., Kaneko, R., Watanabe, Y. and Kamata, K. (2001) 'Nonylphenol Content in Polyvinyl Chloride Wrapping Film for Food and Migration into Food Samples', *Annual Report of Tokyo Metropolitan Research Laboratory of Public Health* 52: 180–84.

GESAMP (2010) Proceedings of the GESAMP International Workshop on Plastic Particles as a Vector in Transporting Persistent, Bio-accumulating and Toxic Substances in the Oceans, ed. T. Bowmer, P.J. Kershaw, GESAMP Rep. Stud. No. 82, pp. 68.

Heskett, M., Takada, H., Yamashita, R., Yuyama, M., Ito, M., *et al.* (2012) 'Measurement of Persistent Organic Pollutants (POPs) in Plastic Resin Pellets from Remote Islands: Toward Establishment of Background Concentrations for International Pellet Watch', *Marine Pollution Bulletin* 64: 445–48.

Hirai, H., Takada, H., Ogata, Y., Yamashita, R., Mizukawa, K., *et al.* (2011) 'Organic Micropollutants in Marine Plastics Debris from the Open Ocean and Remote and Urban Beaches', *Marine Pollution Bulletin* 62: 1683–92.

Isobe, T., Nakada, N., Mato, Y., Nishiyama, H., Kumata, H. and Takada, H. (2002) 'Determination of Nonylphenol Migrated from Food-contact Plastics', *Journal of Environmental Chemistry* 12: 621–25.

Karapanagioti, H.K., Ogata, Y. and Takada, H. (2010) 'Eroded Plastic Pellets as Monitoring Tools for Polycyclic Aromatic Hydrocarbons (PAH): Laboratory and Field Studies', *Global Nest Journal* 12: 327–34.

Kawanaka, Y., Torigai, M., Yun, S.-J., Hashita, T. and Iwashima, K. (2000) 'Elution of Nonylphenols and Bisphenol-A from Plastic Films for Food-wrapping', *Journal of Environmental Chemistry* 10: 73–78.

Kwan, C.S., Takada, H., Mizukawa, K., Hagino, Y., Torii, M., *et al.* (n.d.) 'PBDEs in Leachates from Municipal Solid Waste Dumping Sites in Tropical Asian Countries: Phase Distribution and Debromination', *Environmental Science and Pollution Research*, forthcoming.

Law, K.L., Moret-Ferguson, S., Maximenko, N.A., Proskurowski, G., Peacock, E., *et al.* (2010) 'Plastic Accumulation in the North Atlantic Subtropical Gyre', *Science* 329: 1185–88.

Mato, Y., Isobe, T., Takada, H., Kanehiro, H., Ohtake, C. and Kaminuma, T. (2001) 'Plastic Resin Pellets as a Transport Medium for Toxic Chemicals in the Marine Environment', *Environmental Science and Technology* 35: 318–24.

Ogata, Y., Takada, H., Mizukawa, K., Hirai, H., Iwasa, S., *et al.* (2009) 'International Pellet Watch: Global Monitoring of Persistent Organic Pollutants (POPs) in Coastal Waters. 1. Initial Phase Data on PCBs, DDTs, and HCHs', *Marine Pollution Bulletin* 58: 1437–46.

Moore, C.J., Moore, S.L., Leecaster, M.K. and Weisberg, S.B. (2001) 'A Comparison of Plastic and Plankton in the North Pacific Central Gyre', *Marine Pollution Bulletin* 42: 1297–300.

Ryan, P.G., Connel, A.D. and Gardner, B.D. (1988) 'Plastic Ingestion and PCBs in Seabirds: Is There a Relationship?' *Marine Pollution Bulletin* 19: 174–76.

Schulte-Oehlmann, U., Tillmann, M., Casey, D., Duft, M., Markert, B. and Oehlmann, J. (2001) 'Xeon-estrogenic Effects of Bisphenol-A in Prosobranchs (Mollusca: Gaastropoda: prosbranchia), Z. Umweltchem', *Okotox* 13: 319–33.

Schuur, A.G., Legger, F.F., van Meeteren, M.E., Moonen, M.J.H., van Leeuwen-Bol, I., *et al.* (1998) 'In Vitro Inhibition of Thyroid Hormone Sulfation by Hydroxylated Metabolites of Halogenated Aromatic Hydrocarbons', *Chemical Research in Toxicology* 11: 1075–81.

Takada, H. (2006) 'Call for Pellets! International Pellet Watch Global Monitoring of POPs Using Beached Plastic Resin Pellets', *Marine Pollution Bulletin* 52: 1547–48.

Tanaka, K., Takada, H., Yamashita, R., Mizukawa, K., Fukuwaka, M. and Watanuki, Y. (n.d.) 'Accumulation of Plastic-derived Chemicals in Tissues of Seabirds Ingesting Marine Plastics', *Marine Pollution Bulletin*, forthcoming.

Teuten, E.L., Saquing, J.M., Knappe, D.R.U., Barlaz, M.A., Jonsson, S., *et al.* (2009) 'Transport and Release of Chemicals from Plastics to the Environment and to Wildlife', *Philosophical Transactions of the Royal Society B: Biological Sciences* 364 (1526): 2027–45.

Toyooka, T. and Oshige, Y. (2000) 'Determination of Alkylphenols in Mineral Water Contained in PET Bottles by Liquid Chromatography with Coulometric Detection', *Analytical Sciences* 16: 1071–76.

United Nations Environment Programme (UNEP) (2011) *UNEP Year Book 2011: Emerging Issues in Our Global Environment*, Nairobi: UNEP, 21–33, www.unep.org/yearbook/2011/pdfs/UNEP_YEARBOOK_Fullreport.pdf.

Wintgens, T., Gallenkemper, M. and Melin, T. (2003) 'Occurrence and Removal of Endocrine Disrupters in Landfill Leachate Treatment Plants', *Water Science and Technology* 48: 127–34.

Yamashita, R., Takada, H., Fukuwaka, M.-A. and Watanuki, Y. (2011) 'Physical and Chemical Effects of Ingested Plastic Debris on Short-tailed Shearwaters, *Puffinus tenuirostris*, in the North Pacific Ocean', *Marine Pollution Bulletin* 62: 2845–49.

12 Plastic and the work of the biodegradable

Jennifer Gabrys

The seas and oceans have become a slurry of plastic. There are now estimated to be up to 100 million tons of debris in the five gyres where plastic debris collects in still ocean currents from the Pacific to the North Atlantic.[1] However, the plastic found in these gyres and suspended throughout the seas is not exclusively composed of identifiable objects in the form of water bottles, toy ducks or sandwich bags, but also consists of microplastics. These small-scale pellets, or nurdles, and other plastic fragments are residues from the breakdown of plastic products or fallout from manufacturing sites where tiny plastic feedstock drifts in considerable quantities from factory lots to the seas. Plastics are materials in process; they fragment and break down, while also generating new material arrangements. In what ways do the plastics that are accumulating in oceans give rise to new environmental processes? Who or what are the agents involved in working through the new materialities and effects of plastics as they accumulate and break down in the earth's oceans?

Material processes of accumulation and biodegradability have become evident in many different modes of working through plastics. For example, the amassing of plastics in seas and oceans has given rise to new ways of working through plastics, such as the recent European Union (EU) Maritime Affairs and Fisheries initiative to pay fishermen in the Mediterranean to catch plastic rather than fish (Damanaki 2011). On the one hand, this initiative addresses the problem of over-fishing and disposal of less desirable fish for market, but, on the other hand, it demonstrates that the seas and oceans are now a shifting, if distinctly plastic, material matrix of chemical-biotic-economic processes. Fishing for plastics, it may turn out, could be an economic alternative to fishing for fish, since plastics may be retrieved year round, and the demand for (recycled) plastics feedstock continues to rise.

Fishing for plastics also seems to address the pollution of the seas, which not only affects water quality but also impairs the lives of many marine organisms. Images of dead seabirds that have starved from a stomach full of plastic, together with tales of fish and turtles who 'mistake' plastic for food, and through ingesting this debris eventually die, are regular features of scientific and public concern (Moore *et al.* 2001; Barnes *et al.* 2009). At the same time, newly identified forms of microbial life appear to be emerging that ingest plastic in the seas – although to what effect is yet to be determined, since it is likely that these bacterial forms of ingesting and decomposing plastics also release chemicals for distribution in the seas and concentration in food chains (Zaikab 2011). Yet these labouring bacteria seem to offer an ideal image of how the seas might be cleaned of our offending debris, in the ever-elusive search to eliminate the negative effects of plastics.

In each of these examples, new encounters, practices and natures emerge through material entanglements with plastics. Accumulation in this sense points less towards an exclusive emphasis on environmental contamination and more towards processes of environmental modification in which we are situated with multiple more-than-human entities. It may seem that one way to deal with plastics accumulating in oceans is to fish them out and remove them from the seas. Yet plastics are accumulating in many different ways, as they break down, enter food chains as plasticizers and generate alterations in the eating patterns of diverse organisms. How might a material politics of plastics that is less inclined toward, a purifying discourse of environments and that is more invested in attending to the emergence of new material arrangements make possible a greater engagement with the new natures and practices to which we are committing ourselves – and more-than-humans? How do the new entities and processes that emerge in plasticized oceans shift our understandings and approaches to the material-political ecologies of these spaces?

Accumulation here refers not just to the literal accretion of residual matter in the seas, but also to the build-up of plastics within environmental processes and corporealities. Such 'natures-in-the-making' as well as 'bodies-in-the-making', as Harvey and Haraway (1995: 514) suggest, are junctures where political possibilities may emerge in relation to new material processes and arrangements. Material politics, in this sense, describes the ways in which the materialities we are involved in making are sites not just of responsibility and concern, but also of ongoing – if often problematic – invention. As Thompson's and Takada's chapters in this collection demonstrate, there are numerous new effects and entities emerging with the ongoing presence of plastics in environments. From marine organisms that ingest plastics with concentrated levels of persistent organic pollutants (POPs), to bacteria and algae that colonize plastic, and marine organisms that incorporate plastic debris as habitat or flotation medium, plastics are having considerable effects on organisms and environments. This chapter then discusses how the accumulation of plastics in oceans gives rise to new natures and bodies in the making, as well as new modes of working through these material arrangements.

In order to take up these multiple and different ways in which plastics are accumulating across environments and bodies, I mobilize a notion of material politics that attends to how plastics are entangled with and generative of specific forms of more-than-human *work*. The notion of work is important for this investigation because it allows for an approach to plastics that accounts for the complex creaturely and environmental processes that coalesce in relation to these materials, as well as the political possibilities that might emerge through these natures and bodies in the making. The making of bodies and natures involves the relational 'work' of bodies as they 'hold sites together' (Woodward *et al.* 2010: 274). But the processes whereby sites hold together also change, and so a shifting range of heterogeneous entities undertake material practices that specifically concresce in the actual occasions of plastics as they degrade in the oceans. Drawing on Whitehead (1929) in this understanding of material processes, I suggest here that the ways in which plastics are encountered and worked through as historical forms sedimented in the present also inform the future potential processes that may be undertaken in relation to plastics.

The concept and practice of work and working materialities points to ways in which it may further be possible to reconceptualize the notion of 'carbon workers', a term that refers to the diverse – if at times problematic – ways in which any number of humans and more-than-humans are enrolled in the work of mitigating climate change.[2] Here, I extend and translate this notion of carbon workers towards plastics. Plastics are composites of carbon, both in their physical form as petrochemical hydrocarbons and in the carbon energy used to manufacture them. Eight per cent of world oil production contributes to the substance and energy required to manufacture plastics (Thompson *et al.* 2009b). As composites of carbon, plastics are participants in and mobilize distinct types of carbon work, particularly at end-of-life. Plastics accumulate, break down and degrade, but these processes also enrol humans and more-than-humans in different forms of carbon work. Upon disposal, plastics travel to those carbon sinks of oceans and landfills. In these zones, they further degrade and, depending upon chemical composition, may release carbon dioxide or lodge in the bodies of ocean organisms, thereby diversely influencing the material composition of the ocean as a carbon sink (or source).

I focus on *biodegradability* as a specific form of carbon work that involves processes of transformation, deformation and generation of materials and bodies. Biodegradability has at times been a sought-after quality for plastics, as it signals the seamless elimination of this highly disposable material. Most plastics do not actually biodegrade, but instead *degrade* into smaller particles through chemical processes and physical weathering. Numerous environmental, chemical and biological impacts occur along with these degradation processes across organisms. The actual and typically problematic ways in which plastics do break down – by adsorbing chemicals, entering food chains, and altering biological and reproductive processes through increased levels of toxicity – indicate how degradation and biodegradation are as much political as ecological processes that inform the possibilities of natures and bodies in the making. Biodegradation may be the sought-after quality for plastics, but degradation is the concrete way in which plastics dematerialize and rematerialize to generate new environmental conditions. Even when plastics do biodegrade, they often do not completely disappear but instead fragment into smaller invisible pieces. The *bio*-of degradation then has as much to do with the forms of life – the organisms, processes and environments – that are drawn into the ongoing breakdown of plastics, whether by inadvertently ingesting microplastics or undergoing increased exposure to pollutants that are concentrated on plastic debris surfaces.

The material-political dimensions of biodegradation become more evident through the notion of carbon workers, which is a way to capture the active, material, productive and participative ways in which humans and more-than-humans work through and remake plastics and plastic environments. How might the multiple ways in which plastics are worked through begin to give rise to a material politics of plastics that accounts for these more-than-human modes of carbon work? What types of carbon work become identifiable in relation to plastics as they biodegrade, and what potential types of work might emerge to generate new material political practices?

Accumulation

The plastics accumulating in seas have been storing up and breaking down since the post-World War II rise in plastic consumer goods (Ryan *et al.* 2009). Plastics in the seas are now present in considerable densities, a record of accumulation that is due in many ways to the increasing quantities of plastics, since as Richard Thompson and colleagues (Thompson *et al.* 2009b: 2154) write 'the production of plastic has increased substantially over the last 60 years, from around [half a] million tons in 1950 to over 260 million tons today'. Plastics also collect and sediment over time in cumulative quantities. All plastics ever manufactured since the rise of the Plastic Age are still likely to be present in the environment and oceans in some form, as they will not have completely broken down yet (Lebwohl 2010; Andrady 2003).

While the oceans were relatively free of plastics prior to the post-war Plastic Age, now they are a pervasive substance circulating through oceans, and could even be considered a common entity within ocean ecologies. Oceans are becoming new material compositions, as literary scholar Patricia Yaeger suggests, since with plastic accumulation 'we've reconstituted the physical ocean in a mere fifty years' (Yaeger 2010: 538). In this era of the Anthropocene, not just atmospheres but also oceans are part of ongoing environmental alterations. The reconstitution of the oceans refers less to an essential originary nature, rather indicating how the new natures now emerging are spatial and temporal accumulations of lived materialities. In this plasticization of the ocean, our present and future material politics are then necessarily committed to responding to these natures-in-the-making.

Just as plastic accumulation is taking place in oceanic sinks, these sinks then become spaces where complex biochemical and environmental 'intra-actions' occur across microbial, vegetable and animal corporealities (Barad 2003: 810). Intra-action, as Karen Barad explains, describes processes where entities can be seen to emerge through – rather than prior to – relations.

Bodies and natures form *in* and *through* shared contexts. In the space of plastic accumulation, both humans and more-than-humans take part in material and relational exchanges filtered through plastics and their residues.

Such intra-actions take many forms. Plastic debris is now a frequent transport medium for organisms that travel ocean currents. By 'hitchhiking' on fishing gear and disposable takeaway containers, typically invasive species are able to make far-flung journeys on this readily available debris. While in transit, these species are able to reshape places, as they circulate on plastic media to settle into – or 'colonize' – new environments (Gregory 2009). At the same time, plastics have been shown to be an adsorption medium for potentially harmful chemicals, carrying and dispersing additives and plasticizers such as flame retardants, Bisphenol-A (BPA) and phthalates, as well as drawing in and concentrating chemicals from seawater (Song *et al.* 2009; Takada 2013; Thomas *et al.* 2010). When ingested, these plastics then potentially pass on chemical loads to other types of marine life, which regularly make a meal of plastic particles, thereby amplifying chemical effects in the food chain. The intra-actions that occur through plastics are typically pernicious exchanges, where bodies exposed to plastics and plasticizers accumulate plastic effects, and undergo endocrine disruption or physical blockage, as discussed by Thompson and Takada in Chapters 9 and 11, respectively, of this collection.

While accumulation is often read primarily as a Marxian term that describes strategies of property and capital acquisition – and indeed the ocean can be seen as a space of capital accumulation, as the artist Alan Sekula (1995) makes clear in his work – the accumulation of plastics in the oceans demonstrates the more residual effects of these political economic practices. Here, accumulation extends to bodies and environments as sites of 'production' that require working through the residual materialities of plastics. Harvey and Haraway, together and individually, suggest the ways in which bodies and economies that are jointly formed might be called 'corporealization' (Harvey and Haraway 1995: 510; see also Haraway 2007; Harvey 1998), where the body also constitutes an 'accumulation strategy' along with economic modes of accumulation (Harvey 1998). However, with residual plastics, the ways in which political economies materialize may occur long after cycles of production and consumption are complete. Within these residual materialities, multiple participants are involved in distinct and often intra-active practices of working through the accumulation and degradation of plastics. Plastics do not simply break down in ocean environments; rather, they enrol humans and more-than-humans in new processes and practices of working through and with these natures in the making.

Carbon workers

To say that the oceans are polluted with plastics is an approach to environments through contamination that may not fully account for the more-than-human ways of working through plastics that are already taking place. Instead, from the perspective of natures and bodies in the making, accumulating plastics generate specific material conditions within and through which humans and more-than-humans participate, whether through changing the composition of food chains or increasing levels of toxicity in environments. The accumulations and biodegradations of plastics are events that signal the need to open up approaches to plastics through a more-than-human material politics, since the multiple entities affected – and emerging – through these plastic processes involve numerous other actants.

The material politics under consideration here draws on questions recently raised in relation to 'political matter' – namely, how do politics change when more-than-humans enter into these deliberations (Braun and Whatmore 2010)? How do more-than-humans, as integral to material processes, alter practices and understandings of politics (Haraway 2007; Stengers 2010)? Even more than attending to the ways in which more-than-human entities participate in politics, I am interested in specifying how particular material entities and practices emerge as newly relevant contributors to the politics of changing environments. In this respect, I adapt the term 'carbon workers', which has emerged within specific policies to address the human (and more-than-human) contributions to mitigating climate change, to describe the ways in which plastics are worked through, and the material politics that emerge within these specific processes of degradation and biodegradation.

Within climate change discourse, the concept of carbon workers has gained traction to describe the long list of 'tree planters and tenders, measurement technicians, landscape deforestation modellers, carbon accountants, carbon certifiers and verifiers and others' who have emerged to take care of trees and forests that have been identified as key sites of biotic carbon sequestration through the Kyoto Protocol (Fogel 2002: 182; Lövbrand and Stripple 2006: 235). Carbon workers within climate change discourses primarily describe the various roles that humans play in relation to carbon sink policy instruments, and the often problematic matrix of relations that occurs when developed countries seek to offset carbon emissions – where, for instance, indigenous forest dwellers in developing countries are enrolled in performing carbon work in designated biotic sinks (since many of these sinks are tropical forests). Trees – and the many other more-than-humans that inhabit forests – are also implicitly included as carbon workers in this context, since their participation is gauged in relation to the project of reducing carbon. More-than-humans might then be more explicitly included as workers in the carbon project – entities the participation of which becomes identified as relevant in relation to reducing (or contributing to) carbon emissions.

Oceans are another, less recognized carbon sink, since most carbon work has been configured in relation to terrestrial and atmospheric spaces. Yet oceans are also sites of considerable carbon work, and are now beginning to be addressed not just for their absorption of carbon dioxide, but also for the ways in which the biotic–chemical exchanges that take place there are now of interest for 'managing' this other carbon sink (Stone 2010). The accumulation of plastics and plastic additives is one aspect of this project of attending to the oceans, and gives rise to new forms of possible carbon work.

Carbon work is a way to specify particular types of exchanges and practices that take place in relation to plastics accumulating in oceans. By specifying practices – or 'arrangements of practices', as Haraway suggests – 'heterogeneously complex' modes of agency may become more readily apparent as being interwoven with and generative of concrete political occasions and effects (Harvey and Haraway 1995: 520). Carbon work, as discussed here, could be one way to begin to develop a precise attention to the connections and processes within oceanic sinks. Carbon work is also a way to specify the intra-actions that take place in relation to biodegrading plastics in oceans. The examples of the different modes of working through plastics with which I began this chapter signal types of carbon work that variously 'clean up' or break down plastic hydrocarbons. From EU fishers paid to fish for plastic, to marine researchers focused on documenting the effects of degradable plastics, to nongovernmental organizations (NGOs) focused on raising public awareness around plastic pollution, to animals and birds that ingest plastic debris, to bacteria that may biodegrade these materials, a whole range of carbon workers, relations and practices begin to materialize in distinct ways in relation to plastic oceans.

In these processes of accumulating plastic hydrocarbons, the carbon work of humans and more-than-humans articulates distinct material-political relations to the seas. These material intra-actions within plastic oceans are part of what enables processes of materialization to even turn up as carbon work: plastic fragments turn up by accumulating over time in oceans, bodies and seas, and then become the object of clean-up campaigns or toxicity studies. Dead animals turn up: their inability to process plastic through ingestion makes them a visible remainder and reminder of the intractable accumulation of plastic debris and its ongoing effect on biodiversity. Plastic-loving bacteria turn up, inhabiting and apparently decomposing plastic: are they new, or have they been here all along, and could they clean the oceans of excess debris?

New types of carbon work then emerge as possible strategies for dealing with these fragments. Describing these material processes as *carbon work* draws attention to the complex transformations and exchanges within plastic production, consumption and disposal, which involve more-than-humans in our material lives. The material politics of oceans as sinks, and their role within environmental change, make these processes more evident as forms of work, and demonstrate how our material lives are forceful conjugations and sites of material-political engagement, responsibility and invention.

More-than-humans working

It would be possible here to make a long list of all the working animals to be found in more-than-human research, from the research labour of laboratory animals to military dolphins searching for mines and the industrial labour of aquaculture. However, my interest in attending to the work of more-than-humans is less about the direct servicing of animals or other more-than-humans to economic processes, and more about the ways in which new material collectives emerge to do key carbon work in relation to breaking down plastics in oceans. In this sense, the carbon work of plastic-related activity could be seen to be more comparable to what anthropologist Stefan Helmreich (2009) describes in his study, *Alien Ocean*, where 'methane-metabolizing' bacteria at vents in extreme ocean environments consume and exchange methane through a process of chemosynthesis, thereby preventing additional greenhouse gases from entering the atmosphere (Helmreich 2009: 36–37).

Here is an exchange that could be described as a form of work that contributes to attempts to reduce greenhouse gases, which are articulated and monitored across spaces of policy and everyday practice. These bacteria are not immediately a resource, but they do turn up as a more-than-human contribution in the material politics of climate change. The natures that are in the making in this context involve not just changing climates and distributions of greenhouse gases, but also pertain to the ways in which more-than-human processes emerge as relevant or as contributing to specific environmental concerns and actions.

In his metabolic theory of labour and value, Marx excluded the non-human from his definition of human labour. For Marx, labour was an expression of 'man's' metabolic relation with and conversion of 'nature'. Yet this labour is notable not just for its assumed conversion of nature into resource, but also for what it is not. Non-human work does not constitute labour, Marx argues, since 'nature's work' – whether the web of the spider or the hive of the bee – has not undergone a prior mental conception that would, for instance, characterize the labour of an architect conceptualizing a building (Marx 1990: 283–84).[3] The exclusion of non-human work from theories of labour informs the types of material politics that are possible, since non-humans may not then be recognized as participants in our material lives.

If we trouble Marx's assertion of where work might be situated or identified, we can instead consider how a post-Marxian concept of work might not consist of 'man' labouring to transform 'nature' through metabolic relation, but rather occur through intra-actions and processes of materialization that direct new possibilities for material politics. Instead of human-driven metabolic transformations, here we might consider something closer to Michel Serres's *metabolas* as ongoing processes of transformation, where material and environmental exchanges are characterized by all manner of conversions that take place not just in human bodies, but also in 'animals and plants', as well as 'air crystals … cells and atoms' (Serres 1982: 72–73).

Numerous humans and more-than-humans are involved in the process of converting plastics in one way or another, whether it is bacteria breaking down microplastics, seabirds ingesting bottle-tops, or fisherman fishing for plastics. One of the ways in which these carbon works and exchanges might be characterized further is through processes of degradation and biodegradation. Carbon workers are involved in and producers of material exchanges and arrangements that are not so much metabolic and resource driven, but instead necessarily oriented towards realizing new material dynamics and relations in the ongoing attempts to process and break down plastic hydrocarbons. Through these modes of work, new entities emerge, which contribute to a processual reshaping of what counts as material politics.

The work of the biodegradable

The degradation of plastics in oceans and terrestrial environments is part of the contradictory way in which plastics accumulate: not primarily as identifiable objects but mostly in the form of microplastics, chemical migration and bodily accumulation (Guthman 2011; Thompson *et al.* 2009a). As mentioned in the introduction to this chapter, most plastics are not considered *bio*degradable, but rather degrade only in relation to forms of physical weathering (American Chemistry Society 2010), in some cases through exposure to light or oxygen (Thomas *et al.* 2010), or in other cases through the addition of specific 'transition metals' such as iron or cobalt (Cressey 2011; Roy *et al.* 2011). At the same time, these degradable forms of plastics often break down into fragments that last indefinitely in the environment. Even though these plastic fragments are no longer present in an identifiable form, they still persist as debris with toxic effect.

The persistence of plastics for potentially several hundred years (since degradation depends in part upon context) has often served as one of their least redeeming features. Biodegradable plastics, or bioplastics, have been developed in an attempt to find a remedy for the material persistence and recalcitrance of plastics.[4] Rather than having crude oil as their primary substrate, biodegradable plastics are usually made from starch and cellulose – what otherwise are referred to as 'renewable' materials. Since these materials are derived from plants, and may be composted or degraded through anaerobic digestion rather than put into landfill, they are seen as a possible way to address the accumulation of plastics in environments (Song *et al.* 2009: 2127). When biodegradable plastics break down, they decompose into 'carbon dioxide, methane, water, inorganic compounds, or biomass in which the predominant mechanism is the enzymatic action of microorganisms' (Song *et al.* 2009: 2127–28). In order to meet the terms of biodegradability, microorganisms must also completely use up plastic fragments within a set period of time.

Biodegradation presents an ideal vision of matter, lapsing back into 'nature' without leaving a visible residue. To be biodegradable is to be eco-friendly, to embody the promise to disappear into the earth without a trace. Biodegradability – even if this process involves fragmenting into toxic particles – may be seen to be preferable to being confronted with the visual evidence of enduring plastic remains. Biodegradation could be described as a form of biomimesis, where materials 'mimic' the assumed 'natural' tendency of materials towards reintegration into trophic cycles. Yet biomimesis, as Bensaude Vincent articulates, often involves mapping a teleological agenda on to so-called natural processes in order to realize 'economical rationality' in relation to 'natural systems' (Bensaude Vincent 2007, 2011). Biodegradation could then be seen as a way to attempt to naturalize plastic materials so that they seem to spring easily from and return to nature. Yet this also could be seen as a way to elide, if not idealize, the material politics and processes of which plastics are constitutive.

Within the work of biodegradability, moreover, microorganisms provoke alternative conceptions of what material transformations involve – not as a process of becoming invisible, but rather as an articulation of new collectives brought into the space of material politics. What counts as carbon work here does not then reduce down to a singular entity labouring away at a piece of plastic, but instead requires collective environmental conditions and entities – from light and oxygen to microbial life – to come together in the process of plastic degradation.

Bacteria redux

The microplastics that are present in increasing numbers in oceans are often described as having transformed oceans into a 'plastic soup' (de Vrees 2010). Plastic soup indicates not identifiable items for retrieval but more of a turbid medium of plastic deformation. Perhaps in contrast to an image of garbage patches or marine litter as a thick surface layer of bottles and trash bags choking the upper ocean, here instead is an extensive suspended medium of plastic debris and pellets, which variously pass through the bodies of marine life and undergo bacterial transformation. This plastic soup is a site of continual metamorphosis and intra-actions, so that new or previously unrecognized corporeal relations emerge in the newly constituted spaces of the oceans.

In these spaces, biodegradable as well as petroleum-based plastics have been found to undergo the work of 'plastic munchers', or bacteria that are colonizing and may potentially be digesting plastic. The 'discovery' of bacteria that may be consuming plastic has led to the further proposal to find ways to deploy specific microbes on plastic patches in an attempt to clear these spaces of their accumulated residue. Yet the effects of these carbon-working bacteria are yet to be fully understood: to what extent do the bacteria recirculate the chemical effects of plasticizers into the water and through the food chain? How does this process of 'bioremediation' unfold, which other organisms might be affected, and what time spans and resources might it require? One researcher likened the scale of the bacteria's task in consuming garbage patches to one person having to eat the whole of Canary Wharf (BBC News 2010).

While the scale comparison between bacterial decomposition and dense urban districts might appear daunting, a 1970s science-fiction novel, *Mutant 59: The Plastic Eater* (Pedler and Davis 1971), imagines a scene where bacteria capable of biodegrading plastic run amok in London. Due to their reproductive success, the plastic-loving bacteria are able to multiply, chew through and dissolve entire plastic urban infrastructures. From the failure of electrical wires and cables that are insulated with plastic, to the explosion of water pipes that are similarly made of plastic, the indiscriminate appetites of these bacteria are a force that reshapes and shuts down entire cities. As the Introduction and numerous contributions to this edited collection demonstrate, our material lives increasingly are composed of plastics. Because of the extent of plastic materialities, plastic-digesting bacteria could become architectural agents, remaking the pervasive plastic fabric of our environments. In *Mutant 59*, the urban environment becomes an apocalyptic experiment in degradation, where these new bacterial forms develop evolving appetites and capacities for material transformation as they eat and alter plastic scenes.

Biodegradability and 'eating well'

The work of plastics biodegradation is thus less about making the effects of ongoing disposal-oriented consumerism disappear, since even degradation and biodegradation generate new intra-actions and material politics. Instead, biodegradability points to how the residual materialities of plastics activate a more collective understanding of material processes. As sociologist Myra Hird (2010) suggests, material processes may be indicative of 'eating well', since bacteria are the fixers or producers that make available the elements on which so many heterotrophs, or organisms that require external nourishment, depend. Eating well, she suggests, drawing on Jacques Derrida (1991), is a way to encounter bacteria (and processes such as decomposition) as part of the material collectives in which we all participate. These exchanges and relations might also give rise to indigestion, as Haraway (2007) suggests, or to modes of exchange and incorporation that instead unsettle or disrupt relations.

As I have previously argued in my work on electronic waste and carbon sinks (Gabrys 2009, 2011), in a waste-based materiality, 'things' are rarely present as discrete entities, since materiality involves processes of breaking down, transforming, dispersing and reworking. Hird addresses this lack of discreteness through bacteria in order to articulate how the edges of more-than-humans are not distinct, and how our material processes and politics are always undertaken in collectives. These collectives are sites of ethical relation and obligation. Eating well is about recognizing connections and interdependencies, as well as acknowledging that many more-than-human processes fall outside the scope of our usual sites of recognition. Material collectives are not just sites of eating together, but also of transforming, making possible and available different versions of shared as well as differently inhabited materialities.

Derrida has given us his thoughts on 'eating well', as well as 'biodegradability', which together perhaps offer a revised metabolic imaginary beyond that linear sequence articulated through a more Marxian political (and material) economy. Through the 'figure of the word "biodegradable"', which Derrida transposes to 'cultural uses', he asks: 'What is a thing? What remains? What, after all, of the remains ...?' (Derrida 1989: 812).[5] Biodegradability draws attention to the ways in which things become 'non-things'. Derrida's analysis deals primarily with cultural investigations, but the ways in which things become non-things in the plastic oceans involve multiple material collectives that are undertaking these transformations. In the shifting composition of oceans, bacteria, marine life and fishermen working through EU directives, the carbon work of processing residual plastics in oceans gives rise to newly emerging collective material politics.

Conclusion: material collectives

What types of material politics and material collectives emerge through speculative, expanded and more-than-human modes of carbon work? What might this work consist of, specifically as reconfigured through degradability? Such an approach could be seen to open into all kinds of directions, but I would like to end by discussing how this view of carbon workers and biodegradability as a material-political engagement might open up new types of material thinking. The notion of the 'life-cycle' is a typical device used in the eco-design of products and buildings. It articulates ways for materials to loop back through cycles of production and consumption without material loss or waste. However, from the view of biodegrading and degrading plastics, a life-cycle becomes a very different type of process, far from a closed loop, since the site of recovery might even become a new site of manufacture and material process encompassing the carbon work of multiple material collectives.

Following on from the examples of accumulating, degrading and working through plastics with which I began this chapter in the form of fishers fishing for plastics and bacteria emerging to decompose it, I would like to end with a discussion of one speculative creative practice project, *The Sea Chair Project*, which has been developed to address the increasing amounts of plastics in oceans, and which offers an alternative to life-cycle thinking (Groves *et al.* 2011). The creators of *The Sea Chair Project*, Alexander Groves, Azusa Murakami and Kieren Jones, develop their project as a response to the increasing amounts of plastic found in the seas and at the littoral margins. The project participants have developed a 'nurdler' device, a 'sluice-like contraption' for collecting and sorting plastic debris and microplastic pellets from the ocean. Working in the first instance at the strandline of Porthtowan beach in Cornwall, Grove, Murakami and Jones salvaged plastics for reuse at this site where a particularly large amount of plastic debris and pellets collects. Salvaged plastics were separated by density (and colour) through a floatation-tank technique. The material was then heated in a 'sea press', a furnace and hydraulic press that may be transported on small fishing vessels. The mouldable plastic material was then shaped into a chair – or, more precisely, a three-legged stool (see Figures 12.1–12.5).

In this speculative materials-reclamation proposal, the project creators are specifically interested in addressing the EU initiative to have fishers catch plastic (mentioned throughout this chapter). Here, they have taken this proposal further by developing plans for a sort of 'floating factory ship' that salvages plastic for the production of sea chairs. Rather than work towards an ideal closed-loop life-cycle product, *The Sea Chair Project* works with those historical remains of our lived plastic materialities to begin to generate new approaches to how plastics orient material practices and politics in the present. The reclamation of plastics from oceans is not a straightforward solution to increasing amounts of plastics in oceans, since any project that collects plastics, particularly microplastics, must also attend to the numerous marine (micro)organisms that may be caught up with any collection effort. However, the shifting material arrangements of plastics in oceans here give rise to speculative practices for salvaging degrading plastics as a resource for renewed production. New rounds of production turn from sourcing raw or even recycled materials made raw again, towards the ongoing – if problematic – accumulations of plastics in oceans. Far from a closed loop, the site of

Figure 12.1 'The Nurdler' (2011), *The Sea Chair Project*
Source: (Groves, Murakami and Jones 2011)

Figure 12.2 'Nurdle Collection' (2011), *The Sea Chair Project*
Source: (Groves, Murakami and Jones 2011)

recovery becomes a new site of manufacture and material process. The *carbon work* that emerges does not consist of closed loops of original material recycled again; instead, it generates transformed practices, intra-actions, economies, ecologies and material politics in relation to plastic oceans.

On the one hand, it is sensible – as many researchers have suggested – to deal with the problem of plastics contaminating oceans at the source, to strive either for a policy of minimal waste through redesign or to ensure that plastics do not travel, whether through wayward manufacturing or disposal, to seas. On the other hand, though, the current permeation of oceans and environments with plastics and their chemical residues suggests additional approaches to plastic waste as it already exists are also relevant. Large quantities of plastics continue to be generated and disposed of across established and emerging economies. Many of these economies currently lack waste-handling infrastructures and manufacturing practices that would capture plastic waste before it enters the environment. Hawkins (2010) suggests that it is useful to attend to the ways in which particular materialities may become manifest through environmental practices.[6] Specific materialities may be activated in the actions of banning bags, for instance, or through the uncanny reuse of these same items. These specific materialities are also the sites where 'political capabilities' emerge (Hawkins 2010: 46). By attending to the ways in which materialities are constituted, sustained and produced, it is also possible to consider what practices might prompt alternative forms of material politics. By rethinking the material collectives and material

Figure 12.3 'Plastic Sample: Black' (2011), *The Sea Chair Project*
Source: (Groves, Murakami and Jones 2011)

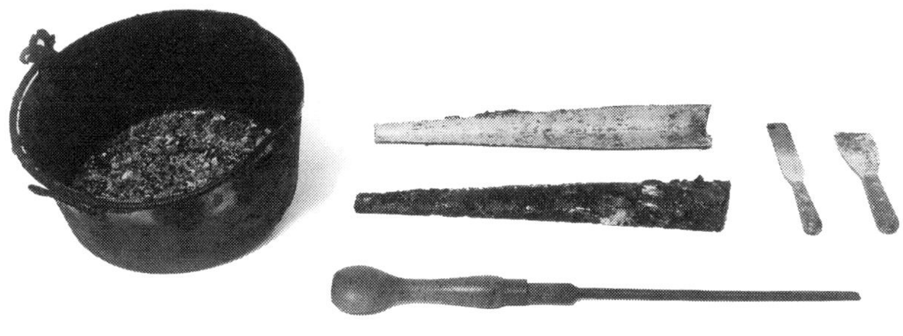

Figure 12.4 'The Sea Chair Tools' (2011), *The Sea Chair Project*
Source: (Groves, Murakami and Jones 2011)

Figure 12.5 'The Sea Chair' (2011), *The Sea Chair Project*
Source: (Groves, Murakami and Jones 2011)

politics that are emerging in relation to plastics in oceans, and the new carbon work to be undertaken there, it may be possible to attend more effectively and more creatively to the material entanglements within which we are now situated.

Such forms of material engagement and material politics perhaps direct us towards what Barad (2010: 266) calls 'an ethics of entanglement', which 'entails possibilities and obligations for reworking the material effects of the past and the future'. Reworking is also a way of working, and carbon reworking engages with and transforms the sedimented effects of environmental and bodily pasts as they turn up in the present and future. Toward what forms of entanglement are we working, and to which natures and bodies in the making are we committed? Which material collectives are brought together, and how do these material relations articulate and make possible different modes of material politics that work through and with these multiple connections?

Materialities and material collectives inform politics, but they are also part of the *becoming possible* of politics. These possibilities of politics are located within forms of work that transform and concretize everyday practices. Bacteria are now establishing their factories in the seas; marine organisms are building highways on polystyrene; and plastic trash is mobilizing human and non-human bodies to work through these oceanic discards in any number of ways. However, these material residues also provide fodder for rethinking the trajectory of our material politics, outside the closed loop of renewed capital, to a more extensive understanding of and speculative approach to the complex and collective carbon work that emerges from our lived plastic materialities.

Notes

1 Numerous reports and organizations document the increasing amounts of plastics in oceans, including the United Nations Environment Programme (2005); and Allsopp *et al.* (2006). The UN report suggests the estimates of plastics per square kilometre should be read with caution, as it is very difficult to gauge exactly how much plastic is in oceans, given how 'vast and varied' they are. For more information on the ocean gyres where plastics collect, see 5 Gyres (n.d.); and US Environmental Protection Agency (2011).

2 The term 'carbon workers' is used across climate-change literature to refer to specific forms of work that emerge in relation to carbon sinks (via their designation in the Kyoto Protocol). Cathleen Fogel (2002, 2004) has addressed this topic briefly in her work on the Kyoto Protocol. Eva Lövbrand and Johannes Stripple (2006) draw on Fogel's work, and briefly deploy the term in relation to understanding the territories of carbon sinks and possibilities for mitigating climate change.

3 In Marx's analysis, transformations of nature are the basis for human labour – but 'nature' also transforms through these processes, and so generates new conditions in which to work.

4 An early report on this varied phenomenon tends towards the science fictional, as in BBC News (1999). For a current industry perspective and overview on bioplastics, see en.european-bioplastics.org (accessed 20 August 2012).

5 While Derrida's text is largely oriented towards a debate on Paul de Man and several academics' interpretations of his work, he deploys the material and

metaphoric language of waste to undertake an analysis of the persistence or dissolution of scholarly work. The use of biodegradables via Derrida is a lateral interpretation, yet his suggestion of how things become non-things is instructive for this study on plastics. As Derrida writes, 'On the one hand, this thing is not a thing, not – as one ordinarily believes things to be – a *natural* thing: in fact, "biodegradable," on the contrary, is generally said of an artificial product, most often an industrial product, whenever it lets itself be decomposed by microorganisms. On the other hand, the "biodegradable" is hardly a thing since it remains a thing that does not remain, an essentially decomposable thing, destined to pass away, to lose its identity as a thing and to become again a non-thing' (Derrida 1989: 813).

6 Hawkins asks: 'How would the politics of plastic bags be understood if the focus shifted from questions of effects to questions of practice?' (Hawkins 2010: 43).

References

5 Gyres (n.d.) 5gyres.org (accessed 19 July 2012).

Algalita Marine Research Foundation (n.d.) www.algalita.org/index.php (accessed 19 July 2012).

Allsopp, M., Walters, A., Santillo, D. and Johnston, P. (2006) *Plastic Debris in the World's Oceans*, Amsterdam: Greenpeace International, www.greenpeace.org/international/en/publications/reports/plastic_ocean_report (accessed 19 July 2012).

American Chemistry Society (2010) 'Hard Plastics Decompose in Oceans, Releasing Endocrine Disruptor BPA', *ACS News Release*, 23 March, portal.acs.org/portal/acs/corg/content?_nfpb=true&_pageLabel=PP_ARTICLEMAIN&node_id=222&content_id=CNBP_024352&use_sec=true&sec_url_var=region1&-uuid=be0e851a-5400-474f-a7cb-311930109 61a (accessed 19 July 2012).

Andrady, A.L. (ed.) (2003) *Plastics and the Environment*, Hoboken, NJ: John Wiley & Sons.

Barad, K. (2003) 'Posthumanist Performativity: Toward an Understanding of How Matter Comes to Matter', *Signs: Journal of Women in Culture and Society* 28(3): 801–31.

——(2010) 'Quantum Entanglements and Hauntological Relations of Inheritance: Dis/continuities, Spacetime Enfoldings, and Justice-to-come', *Derrida Today* 3(2): 240–68.

Barnes, D.K.A., Galgani, F., Thompson, R.C. and Barlaz, M. (2009) 'Accumulation and Fragmentation of Plastic Debris in Global Environments', *Philosophical Transactions of the Royal Society B: Biological Sciences* 364(1526): 1985–98.

BBC News (1999) 'Can the Oceans be Cleared of Floating Plastic Rubbish?' 6 October.

——(2010) 'Scientists Unveil Plastic Plants', 28 September.

Bensaude Vincent, B. (2007) 'Reconfiguring Nature Through Syntheses: From Plastics to Biomimetics', in B. Bensaude-Vincent and W.R. Newman (eds) *The Natural and the Artificial: An Ever-evolving Polarity*, Cambridge, MA: MIT Press.

——(2011) 'A Cultural Perspective in Biomimetics', *Advances in Biomimetics*, 3 February, nano2e.org/?p = 303 (accessed 19 July 2012).

Braun, B. and Whatmore, S. (eds) (2010) *Political Matter: Technoscience, Democracy, and Public Life*, Minneapolis: University of Minnesota Press.

Butler, J. (1993) *Bodies that Matter: On the Discursive Limits of 'Sex'*, New York: Routledge.

Cressey, D. (2011) 'Puzzle Persists for "Degradable" Plastics', *Nature News*, 21 April.

Damanaki, M. (2011) *How Plastic Bags Pollute the Future of Our Seas*, Athens: European Commission for Maritime Affairs and Fisheries.

Derrida, J. (1989) 'Biodegradables: Seven Diary Fragments', *Critical Inquiry* 15(4): 812–73.

——(1991) '"Eating Well" or the Calculation of the Subject: An Interview with Jacques Derrida', in E. Cadava, P. Connor and J.-L. Nancy (eds) *Who Comes After the Subject?* New York: Routledge.

de Vrees, L. (2010) 'Marine Litter: Plastic Soup and More', Brussels: European Commission, Environment Directorate-General, 8 November, ec.europa.eu/environment/water/marine/pdf/report_workshop_litter.pdf (accessed 19 July 2012).

European Bioplastics (n.d.) en.european-bioplastics.org (accessed 19 July 2012).

Fogel, C. (2002) 'Greening the Earth with Trees: Science, Storylines and the Construction of an International Climate Change Institution', unpublished PhD thesis, University of California, Santa Cruz.

——(2004) 'The Local, the Global, and the Kyoto Protocol', in S. Jasanoff and M.L. Martello (eds) *Earthly Politics: Local and Global in Environmental Governance*, Cambridge, MA: MIT Press, 103–25.

Gabrys, J. (2009) 'Sink: The Dirt of Systems', *Environment and Planning D: Society and Space* 27(4): 666–81.

——(2011) *Digital Rubbish: A Natural History of Electronics*, Ann Arbor, MI: University of Michigan Press.

Gregory, M.R. (2009) 'Environmental Implications of Plastic Debris in Marine Settings: Entanglement, Ingestion, Smothering, Hangers-on, Hitch-hiking and Alien Invasions', *Philosophical Transactions of the Royal Society B: Biological Sciences* 364(1526): 2013–25.

Groves, A., Murakami, A. and Jones, K. (2011) *The Sea Chair Project*, www.seachair.com (accessed 20 August 2012).

Guthman, J. (2011) 'Does Eating (Too Much) Make you Fat?' in J. Guthman, *Weighing In: Obesity, Food Justice, and the Limits of Capitalism*, Berkeley, CA: University of California Press, 91–115.

Haraway, D. (2007) *When Species Meet*, Minneapolis, MN: University of Minnesota Press.

Harvey, D. (1998) 'The Body as an Accumulation Strategy', *Environment and Planning D: Society and Space* 16(4): 401–21.

Harvey, D. and Haraway, D. (1995) 'Nature, Politics, and Possibilities: A Debate and Discussion with David Harvey and Donna Haraway', *Environment and Planning D: Society and Space* 13(5): 507–27.

Hawkins, G. (2010) 'Plastic Materialities', in B. Braun and S. Whatmore (eds) *Political Matter: Technoscience, Democracy and Public Life*, Minneapolis, MN: University of Minnesota Press, 119–38.

Helmreich, S. (2009) *Alien Ocean: Anthropological Voyages in Microbial Seas*, Berkeley, CA: University of California Press.

Hird, M.J. (2010) 'Meeting with the Microcosmos', *Environment and Planning D: Society and Space* 28(1): 36–39.

Lebwohl, B. (2010) 'Anthony Andrady says Plastics in Ocean Biodegrade Slowly', *EarthSky*, 11 January, earthsky.org/earth/anthony-andrady-plastics-in-ocean-biodegrade-slowly (accessed 19 July 2012).

Lövbrand, E. and Stripple, J. (2006) 'The Climate as Political Space: On the Territorialisation of the Global Carbon Cycle', *Review of International Studies* 32(2): 217–35.

Marx, K. (1990 [1867]) *Capital: A Critique of Political Economy, Volume 1*, trans. B. Fowkes, Harmondsworth: Penguin.

Moore, C.J., Moore, S.L., Leecaster, M.K. and Weisberg, S.B. (2001) 'A Comparison of Plastic and Plankton in the North Pacific Central Gyre', *Marine Pollution Bulletin* 42(12): 1297–300.

Pedler, K. and Davis, G. (1971) *Mutant 59: The Plastic Eater*, London: Souvenir Press.

Roy, P.K., Hakkarainen, M., Varma, I.K. and Albertsson, A.C. (2011) 'Degradable Polyethylene: Fantasy or Reality?' *Environmental Science and Technology* 45(10): 4217–27.

Ryan, P.G., Moore, C.J., van Franeker, J.A. and Moloney, C.L. (2009) 'Monitoring the Abundance of Plastic Debris in the Marine Environment', *Philosophical Transactions of the Royal Society B: Biological Sciences* 364(1526): 1999–2012.

Sekula, Allan (1995) *Fish Story*, Düsseldorf: Richter Verlag.

Serres, M. (1982) *Hermes: Literature, Science, Philosophy*, eds J.V. Harari and D.F. Bell, Baltimore, MD: Johns Hopkins University Press.

Song, J.H., Murphy, R.J., Narayan, R. and Davies, G.B.H. (2009) 'Biodegradable and Compostable Alternatives to Conventional Plastics', *Philosophical Transactions of the Royal Society B: Biological Sciences* 364(1526): 2127–39.

Stengers, I. (2010) 'Including Nonhumans in Political Theory: Opening Pandora's Box?' in B. Braun and S. Whatmore (eds) *Political Matter: Technoscience, Democracy, and Public Life*, Minneapolis, MN: University of Minnesota Press, 3–34.

Stone, R. (2010) 'The Invisible Hand Behind a Vast Carbon Reservoir', *Science* 328 (5985): 1476–77.

Takada, H. (2013) 'International Pellet Watch: Studies of the Magnitude and Spatial Variation of Chemical Risks Associated with Environmental Plastics', in J. Gabrys, G. Hawkins and M. Michael (eds) *Accumulation: The Material Politics of Plastic*, London: Routledge.

Thomas, N., Clarke, J., McLauchlin, A. and Patrick, S. (2010) *Assessing the Environmental Impacts of Oxo-degradable Plastics Across Their Life Cycle*, London: Department for Environment, Food and Rural Affairs.

Thompson, R.C., Moore, C.J., vom Saal, F.S. and Swan, S.H. (2009a) 'Plastics, the Environment and Human Health: Current Consensus and Future Trends', *Philosophical Transactions of the Royal Society B: Biological Sciences* 364(1526): 2153–66.

Thompson, R.C., Swan, S.H., Moore, C.J. and vom Saal, F.S. (2009b) 'Our Plastic Age', *Philosophical Transactions of the Royal Society B: Biological Sciences* 364 (1526): 1973–76.

United Nations Environment Programme (2005) 'Marine Litter: An Analytical Overview', www.unep.org/regionalseas/marinelitter/publications/docs/anl_oview.pdf (accessed 19 July 2012).

US Environmental Protection Agency (EPA) (2011) *Marine Debris in the North Pacific: A Summary of Existing Information and Identification of Data Gaps*, San Francisco: EPA.

Whitehead, A.N. (1929) *Process and Reality: An Essay in Cosmology*, New York: The Free Press.

Woodward, K., Jones, J.P. III and Marston, S.A. (2010) 'Of Eagles and Flies: Orientations Toward the Site', *Area* 42(3): 271–80.

Yaeger, P. (2010) 'Sea Trash, Dark Pools, and the Tragedy of the Commons', *Publications of the Modern Language Association* 125(3): 523–45.

Zaikab, G.D. (2011) 'Marine Microbes Digest Plastic', *Nature News*, 28 March.

Index

Made in the USA
Middletown, DE
08 September 2021

47825430R00144